ASHRAE Design Guide for Commercial Kitchen Ventilation

Best Practices for Design and Operation

This publication was prepared under the auspices of
ASHRAE Technical Committee (TC) 5.10, Kitchen Ventilation.

Updates and errata for this publication will be posted on the ASHRAE website at www.ashrae.org/publicationupdates.

ASHRAE
Design Guide for Commercial Kitchen Ventilation

Best Practices for Design and Operation

ASHRAE

Peachtree Corners

ISBN 978-1-955516-45-7 (paperback)
ISBN 978-1-955516-46-7 (PDF)

Cover photograph ESB Professional/Shutterstock.com.

Library of Congress Cataloging-in-Publication Data

Names: ASHRAE (Firm), author.
Title: ASHRAE Design Guide for Commercial Kitchen Ventilation : Best Practices for Design and Operation / ASHRAE.
Description: Peachtree Corners : ASHRAE, [2023] | Includes bibliographical references and index. | Summary: "Proper design of a commercial kitchen ventilation system is critical. The design can impact employee safety, fire prevention, energy consumption, building pressurization control, and emission control. This design guide was created with the goal of producing a best practice for design engineers. It was written in accord with current North American codes, standards, and design practices and is intended to aid design engineers in selecting all components comprising a kitchen ventilation system from the initial survey to determine the needs of a system to proper equipment selection, controls and ductwork selection, and final commissioning"-- Provided by publisher.
Identifiers: LCCN 2023007727 | ISBN 9781955516457 (paperback) | ISBN 9781955516464 (adobe pdf)
Subjects: LCSH: Ventilation--Design and construction--Standards--United States. | Kitchens--Design and construction--Standards--United States. | Commercial buildings--Heating and ventilation--Standards--United States.
Classification: LCC TH7684.K5 A84 2023 | DDC 697.95/20973--dc23/eng/20230302
LC record available at https://lccn.loc.gov/2023007727

About the Authors

David A. John, PE, is President of Air Distribution Consultants International and has over 40 years of experience in the HVAC industry. John served as an ASHRAE Distinguished Lecturer from 2010 to 2014 and received the ASHRAE Distinguished Service Award in 2012. He co-chaired the development of the *ASHRAE Design Guide for Air Terminal Units*. John is a former secretary of ASHRAE Technical Committee (TC) 5.10, Kitchen Ventilation; chair of ASHRAE TG 2.RAST, Reactive Air and Surface Treatment; and chair of ASHRAE Standard Project Committee (SPC) 129, Measuring Air-Change Effectiveness. He is a commissioner of the Florida Building Commission and has professional engineer licenses in Texas and Florida.

Frank Kohout, PE, is Director of Engineering for Cyclone Energy Group in Chicago, Illinois. He has over 30 years of HVAC design experience and has served as the lead mechanical engineer for McDonald's Corporation Worldwide Development for 16 of those years. He is currently a member of ASHRAE TC 5.10, Kitchen Ventilation; a voting member of ASHRAE Standing Standard Project Committee (SSPC (SSPC) 154, Ventilation for Commercial Cooking Operations; and a voting member of ASHRAE TC 9.12, Tall Buildings. Kohout holds a professional engineer license in Illinois, ASHRAE's Building Commissioning Professional (BCxP) certification, and LEED AP BD+C credentials.

Derek Schrock is North America Research Director, Foodservice SBA, for Halton Company in Scottsville, Kentucky. He has over 25 years of experience in commercial kitchen ventilation systems and is the current chair of ASHRAE TC 5.10, Kitchen Ventilation.

Russell Robison is the Director of Operations at Gaylord Industries in Tualatin, Oregon, and has nearly 20 years of hands-on commercial kitchen product and applications design achievements. He has served as the chair and vice chair of ASHRAE TC 5.10, Kitchen Ventilation, and has been a contributor to SSPC 154, Ventilation for Commercial Cooking Operations, for 14 years.

Jessica Weaver is the Principal Application Engineer at Accurex, a manufacturer of commercial kitchen ventilation systems and part of the Greenheck Group. Jessica has been involved in mechanical engineering research and design since 2011 and has been specifically designing commercial kitchen ventilation systems since early 2016. She has been actively involved in ASHRAE TC 5.10, Kitchen Ventilation, and SSPC 154, Ventilation for Commercial Cooking Operations, since 2019.

Keith Page is a mechanical engineer who has worked in the chimney/vent/exhaust duct industry for 20 years with Selkirk Corp/Hart & Cooley, Inc. His roles have been in product development, field technical sales support, and product/engineering management. He actively participates on several relevant ASHRAE, NFPA, and UL technical committees.

Mitch Gansen is the National Sales Manager for Schebler Chimney Systems. He has been with the company for seven years, focusing on promoting and educating the engineering and construction industries on chimney/vent/exhaust duct systems. His roles have been in technical sales support, product development, and field coordination and management. Mitch is continually adding to his experience by participating in ASHRAE, NFPA, and UL technical committees.

Greg DuChane is Director of Sales, Trane National Accounts, Trane Technologies and has served at Trane Technologies for more than 33 years. DuChane's experience includes working with large national retail and restaurant chains in various capacities. He serves on the Restaurant Facilities Management Association's online education committee; is a past chair of ASHRAE TC 5.10, Kitchen Ventilation; and is the current chair of the ASHRAE Handbook Committee.

Pictured from left to right: Greg DuChane, Keith Page, Derek Schrock, David John, Russell Robison, Mitch Gansen, Frank Kohout, Jessica Weaver

Contents

Preface

Proper design of a commercial kitchen ventilation system is critical to the proper operation of the system throughout its lifetime. The design can impact employee safety, fire prevention, energy consumption, building pressurization control, and emission control. This design guide was created with the goal of producing the best practices in kitchen ventilation design for engineers. It was written in accordance with current North American codes, standards, and design practices and is intended to aid design engineers in selecting all components composing a kitchen ventilation system, from the initial survey determining the needs of a system to proper equipment selection, controls and ductwork selection, and final commissioning.

ASHRAE Technical Committee (TC) 5.10, Kitchen Ventilation, is the cognizant committee that reviewed and ultimately voted to approve this guide. TC 5.10 is concerned with the design, construction, installation, commissioning, and sustainable operation of code-compliant kitchens.

TC 5.10 is also responsible for the Kitchen Ventilation chapter in *ASHRAE Handbook—HVAC Applications*.

Acknowledgments

The authors would like to acknowledge several individuals who made major contributions to this guide. Jason Greenberg with HFA developed the detailed outline that was used to produce this guide. Jimmy Sandusky with Halton Company contributed to appendix A, "Codes and Standards." Rob Carmichael with Gaylord Industries contributed to chapter 1, "Life-Cycle Cost Analysis." Jason Brown and Bryan Miller, both with the Melink Corporation, contributed to the air balance sections of chapter 8, "TAB and System Commissioning." We would also thank George Stefanovici, President at Carastro & Associates, Inc., for his engineering comments.

The authors also extend appreciation to ASHRAE staff members Cindy Michaels, Mark Owen, and Robyn Weaver for their help and direction in bringing this book to fruition.

Abbreviations

AA	aromatic amine
AABC	Associated Air Balance Council
ACGIH	American Conference of Governmental Industrial Hygienists
ADA	Americans with Disabilities Act
AH	aromatic hydrocarbon
AHJ	authority having jurisdiction
AHU	air handling unit
ANSI	American National Standards Institute
ASTM	ASTM International
BEIs®	Biological Exposure Indices
BMS	building maintenance system
BOD	Basis of Design
C&C	capture and containment
cfm	cubic feet per minute
CKV	commercial kitchen ventilation
CO	carbon monoxide
Cx	commissioning
CxA	commissioning provider
DCKV	demand-controlled kitchen ventilation
DOAS	dedicated outdoor air system
DX	direct expansion
EC	electrically commutated
EER	energy efficiency ratio
EPA	U.S. Environmental Protection Agency
ESP	electrostatic precipitators
FP	fire protection
fpm	feet per minute
GPM	gallons per minute
HEPA	high-efficiency particulate air
HSPF	heating seasonal performance factor
HVAC	heating, ventilation, and air-conditioning
IBC	*International Building Code*
ICC	International Code Council
IECC	*International Energy Conservation Code*
IEER	integrated energy efficiency ratio

IEQ	indoor environmental quality
IMC	*International Mechanical Code*
IR	infrared
IRR	internal rate of return
L/s	litres per second
LCC	life-cycle cost
LEED	Leadership in Energy and Environmental Design®
lf	linear foot
MAU	makeup air unit
MERV	minimum efficiency reporting values
MOT	method of test
MSG	manufacturers' standard gauge
NEBB	National Environmental Balancing Bureau
NEC	*National Electrical Code*
NFPA	National Fire Protection Association
NOx	nitrogen oxides
O&M	operation and maintenance
OPR	Owner's Project Requirements
OSHA	Occupational Safety and Health Administration
PAH	polycyclic aromatic hydrocarbon
PCU	pollution control unit
PM	particulate matter
PMV	predicted mean vote
PRV	power roof ventilator
QSR	quick-service restaurants
rh	relative humidity
RP	research project
RTU	rooftop unit
SMACNA	Sheet Metal and Air Conditioning Contractors' National Association
TAB	testing, adjusting, and balancing
TC	Technical Committee
TLVs®	threshold limit values
UL	Underwriters Laboratories (now UL Solutions)
ULPA	ultra-low penetration air
UMC	*Uniform Mechanical Code*
UV	ultraviolet
UVC	ultraviolet-C
VAV	variable air volume
VFD	variable-frequency drive
VOC	volatile organic compound

Introduction

Why Use This Guide

The purpose of this guide is to provide design engineers, food service consultants, commissioning providers (CxPs), architects, owners, and facility managers with the latest research, information on technology, and resources to design a kitchen ventilation system that meets the owner's requirements while following all recognized best design practices and applicable North American code requirements. It also provides the rationale behind these practices and recommendations. This design guide is not intended to override any local jurisdiction requirements.

One of the primary goals of this guide is to share with the HVAC and food service industry the cumulation of decades of commercial kitchen ventilation (CKV) research and design experience that has transformed the design and operation of CKV systems from rules of thumb into fact-based standards and best practices that are based on scientific research and practical experience.

Another goal of this guide is to show the intent behind its recommendations so other guides outside of North America can be aligned with this one, with the goal of establishing more common best practices and guidelines throughout the world. In addition, future editions of this guide could be aligned with those. The intent of this guide is to

- aid designers in identifying the types of systems and equipment available and evaluating those options to optimize system operation,
- discuss installations and recommendations,
- address operational parameters to consider to ensure the specified size, type of unit, and controls will optimize and harmonize the CKV system's performance, and
- provide ownership expectations (such as comfort, energy savings, etc.), to ensure optimal CKV operation.

Because the design of a CKV system requires incorporating all current best practices, this guide addresses the following:

- Healthy, safe environments in a kitchen
- Energy efficiency and optimization
- Minimizing emissions to the outdoors
- Applicable codes and standards
- Sustainability

The CKV Design Process

Designing a CKV system is a challenging task that requires following a procedure that may not appear intuitive at first glance. To ensure the resulting CKV system operates optimally and as intended, it is recommended that the design sequence be followed in the order the chapters are presented in this book. Note that the first two chapters provide background information that needs to be defined and understood by the designer *before* the actual design work begins. The CKV system design work begins with the exhaust hood selection. The steps below are either covered within the chapters or have their own dedicated chapter (e.g., chapter 3 is dedicated to the first step, hood selection).

A summary of the steps for designing a CKV system is provided here:

- **Hood selection.** Selecting the hood and associated exhaust air volume is always the first task to perform. The hood is selected based on current cooking appliance needs and possible future appliance lineups. This selection includes deciding the type of hood, the configuration of any hood options, and whether any demand ventilation controls are to be used. While hood selection is typically performed by a food service consultant, it is strongly recommended that the engineer be involved with the hood selection process to ensure that the selected hood offers the best engineered value—including safety, performance, first cost, and operating cost—for the client while considering the entire CKV system and not solely the appliance lineup.
- **Exhaust requirements.** It is critical to understand that optimizing the hood's exhaust requirements while maintaining capture and containment (C&C) will minimize the associated replacement air quantity, which will in turn reduce the size of the required HVAC equipment and associated operating costs.
- **Building integration.** Designing a CKV system that integrates into the rest of the building's ventilation design is especially critical if the kitchen is part of a much larger building (i.e., not a free-standing restaurant).
- **Replacement/makeup air.** Determine the sources and quantities of the replacement air/makeup air.
- **System controls.** Determine the CKV system's controls and the associated sequences.
- **Load calculations.** Perform HVAC heating and cooling load calculations.
- **Pollution control units.** Determine the need for pollution control units (i.e., grease, smoke, or odor control).
- **Ventilation.** Design the exhaust system(s) for both cooking effluent and general ventilation (i.e., ductwork and fan selection).
- **Safety.** Design for all related life safety systems, including fire protection.
- **Supply and exhaust air.** Select the supply and exhaust air equipment.
- **Finalization.** Combine all of the above items into a fully designed system.
- **Operations and commissioning.** Determine the operational and commissioning requirements.

This list provides the typical order of the design process for a CKV system. (For specific design guidance, refer to chapter 7, Figure 7.1.) This process should be followed even when redesigning existing CKV systems, because during the redesign, the CKV system design parameters—including appliance lineups, airflow quantities, heat gains to space, exhaust fans, replacement, etc.—may change, which could result in energy savings or an increase in energy consumption depending on the changes.

Designing a CKV system requires a broad array of unique design considerations, including the following:

- The quantity and movement of air as it is delivered into and exhausted from the commercial kitchen space(s). This affects occupant comfort, space pressurization, and all hood exhaust air quantities.
- Accounting for load calculations.
- The sensible, latent, and radiation heat gains from the space's kitchen equipment (or *plug load*, as it is referred to in energy modeling). Kitchens typically require a large amount of cooling for their size, and this must always be calculated—it is never recommended to use a rule of thumb.
- The heat gains and losses from any replacement air, taking into account any fully conditioned, partially conditioned, or unconditioned air.
- Code requirements, including ventilation air requirements (e.g., ANSI/ASHRAE Standard 62.1 [2022c]) and any applicable fire codes. This should include any local requirements for the jurisdiction or from the authority having jurisdiction (AHJ).

Design Guide Chapters

This guide contains the following chapters and appendices, which are outlined below:

- **Chapter 1, "Life-Cycle Cost Analysis."** Many decisions must be made during the CKV design process. Understanding the cost impact—on either the system's first cost or its operating cost—is critical. Given the large amount of energy a CKV system can consume during its normal operation, a minor design decision or investment in up-front costs for energy savings can have a big impact on the utility costs for the life span of the system. Because every CKV design decision should include this type of analysis, this chapter precedes the others.
- **Chapter 2, "Predesign."** This chapter contains questions to ask the owner or for the design team to consider at the project's conception (e.g. type of food service, hours of operation, etc.). The answers to these questions will drive many of the CKV system's design decisions.
- **Chapter 3, "Exhaust Hoods and Related Equipment."** This chapter describes the criteria for selecting an exhaust hood and its associated components for a particular application. While this function is typically performed by the hood supplier, it is critical for the designers (food service and engineer) to understand the selection process, best-practice design, and any associated decisions to better understand the impact of these decisions on the entire CKV system. Doing so provides the knowledge necessary to (1) verify the supplier's recommendations, (2) make any recommended changes to the hood selections, and (3) allow the designers of record to appropriately approve hood selections.
- **Chapter 4, "Design Considerations."** This chapter describes all the details that must be considered when designing a CKV system. Examples include the impact of different cooking processes, replacement/makeup air, and exhaust effluent control.
- **Chapter 5, "Grease Exhaust Duct and Kitchen Hoods."** Given the critical nature of this topic as it relates to building safety and occupant life safety, this chapter is dedicated to the details for designing these systems.
- **Chapter 6, "Occupant Safety and Comfort Considerations."** This chapter describes all safety considerations for designing a CKV system, including air quality and acceptable temperatures and humidity for occupant safety and comfort.
- **Chapter 7, "Comprehensive Design Approach."** This chapter describes the overall CKV HVAC design process and incorporates into that process the critical details presented in previous chapters.

- **Chapter 8, "TAB and System Commissioning."** This chapter describes the critical tasks that must be executed to ensure that kitchens operate properly. A well-designed kitchen will not function as designed if it is not properly air balanced and commissioned. This chapter lists examples and includes critical tasks to verify the air balance has been designed and executed correctly as well as testing and verifications for the proper operation of the CKV system. It also lists important items that should be included as part of the entire CKV system commissioning process, from the project's inception through its conclusion.
- **Appendix A, "Codes and Standards."** This chapter lists the codes and standards that are referenced in this guide or that have CKV-related content.
- **Appendix B, "Research Conducted by ASHRAE TC 5.10."** This chapter lists all CKV-related ASHRAE research projects.

Overall, this guide is intended to be read sequentially. However, it was developed with the understanding that many designers do not have the time to read an entire guide to complete their task at hand. Therefore, it was with this understanding that the chapters of this guide were written so that each chapter can be read by itself for coverage of its topic in detail, and the related topics that are detailed in other chapters are appropriately referenced.

1

Life-Cycle Cost Analysis

Objective

The purpose of this chapter is to assist the designer in understanding the cost of a commercial kitchen ventilation (CKV) system design, both initially and over its lifetime, and using that understanding to meet the owner's expectations and/or help them find a balance between low initial cost and low operational costs so they can make informed decisions for the best application to meet their needs.

Total Cost of Ownership

A well-designed system should include consideration not only of quality equipment but also of the total cost of ownership across the system's lifetime. At the beginning of the project, clear objectives should be established for designers to properly follow to meet design expectations and parameters. These objectives could be existing standards set by the owner or decided upon at the outset of the project by the owner, architect, and engineers. One method engineers can use to justify their design selections is life-cycle cost (LCC) analysis. The LCC evaluation is used when multiple equivalent equipment selections satisfy the design. It takes into consideration not only the initial cost but also operational costs, maintenance costs, efficiencies, repair costs, replacement costs, and system upgrades.

For projects in which high-efficiency buildings, such as buildings being designed to specific Leadership in Energy and Environmental Design® (LEED®) certifications (www.usgbc.org/leed), are the final objective, LCC calculations come to the forefront of equipment considerations. Additional focus should be paid to the LCC process in these instances.

Financial and Economic Guidelines

LCC analysis should be executed with an understanding of the owner's financial requirements and needs, which should frame the equipment selection. Organizations such as hospitals, universities, corporations, and others will have these guidelines in place along with sustainability requirements, hurdle rates, and long-term performance goals. As the business evolves and the facility ages, these guidelines will assist them in their annual budgets beyond year one and over the facility's lifetime.

For those facilities that will be in place for the duration of the LCC time calculation or longer, budgeting is an important aspect of understanding where to allocate dollars and forecast what future costs will look like over the facility's lifetime. Having a reality-based cost analysis gives the owner the tools to make those determinations.

Equivalent Equipment Selection

Measuring equivalent equipment selections is a critical first step in the CKV system design process and gives the engineer a strong basis for LCC calculations. When gauging equivalency, the engineer must consider not only airflows, grades of stainless steel, serviceability, and other parameters but also overall performance and the ventilator's effect on the entire HVAC system.

For HVAC equipment, considerations should be made concerning the size and type of heating and cooling systems depending on geographical location, degree-days, humidity, and other factors or the many types of rooftop units (RTUs) available. These conditions will determine the best fit for the design. For example, evaporative and direct expansion (DX) cooling may both cover the specific cooling load, but the energy required for evaporative cooling will be greater in humid environments. Economizers can save energy in the long term by bringing in outdoor air when appropriate rather than inducing the compressor to start when covering internal loads from machinery, lighting, occupancy, and solar energy becomes necessary. When evaluating long-term costs, high-efficiency units should be considered versus standard-efficiency units.

Other equipment, such as fans and ductwork, also needs to be reviewed when considering total cost of ownership. For fans, high-efficiency motors and variable-speed options are proven ways to save fan energy. Considerations for rooftop versus in-line fans should be made for serviceability and efficiency while also balancing considerations for noise concerns. Once the equipment is sized, the evaluation of ductwork to move the air to the desired location is necessary. Spiral ductwork, listed ductwork, and ductwork built on site are all considerations. For each option, access and serviceability are a concern. When evaluating ductwork systems, it is best to understand what is being brought in or exhausted to determine the best application. While each may deliver or exhaust the desired amount of air, the static losses between these can vary greatly. When exhausting hazardous or combustible material, fire conditions should also be evaluated. Listing agencies have extensive fire testing standards for ductwork, which should be considered.

Intangible items such as comfort, flexibility, and environmental impact may vary by location and be important to a particular facility. Alternatives to the primary design should not only meet the design parameters but also fall into the owner's financial guidelines for minimum rate of return, user preferences, and effect to the overall HVAC system. Decisions made on an individual component life-cycle assessment give only a partial glimpse into the cost analysis, while the sum of the components gives a more holistic view.

Features, Functions, and Benefits

Labor savings. Labor costs are becoming an increasing portion of a facility's operational budget. When evaluating selections, the addition of new or labor-saving features can reduce the total cost of ownership. On projects where local labor rates are high or union labor/prevailing wage is prevalent, it is important to consider whether products with features, functions, and benefits that can reduce labor-intensive duties, thereby reducing maintenance costs, are available. For example, are there systems available for hoods that clean the filters and reduce manual removal, cleaning, and replacement on a daily basis? Water wash systems, high-efficiency filters, ultraviolet (UV) treatments, and pollution control devices are products available to help reduce the amount of grease in the ductwork and on fans, thus reducing the time needed for maintenance. In labor considerations for pollution control, automatic wash functions have long-term cost savings in comparison to the monthly/quarterly removal and replacement of paper filters.

Energy savings. Energy savings are also a concern across the lifetime of a system. Demand-controlled kitchen ventilation (DCKV) features are commonly seen and have a track record of long-term energy savings. When the cooking process is at a reduced state or when cooking equipment is in standby mode, the exhaust fan and dedicated makeup-air fan speeds are subsequently reduced, thus reducing exhaust and supply airflows. Additionally, for the exhaust, lowering the airflow reduces the amount of conditioned air being evacuated while maintaining capture and containment (C&C) of the cooking effluent. The supply airflow for the hood is also reduced, lessening the amount of makeup air entering the space. For this makeup air, it should be noted that less than 50% is recaptured by the hood based on entering temperature, humidity, and space conditions, and at least 50% becomes a load to the space (ASHRAE 2019). This combination allows the HVAC system to work less and cycle on less frequently.

Heat recovery is another tactic to alter a project's LCC. Transferring the heat energy from the cooking process or from heat recovery devices through the HVAC system to other systems requiring heat can also be considered.

Additional savings considerations. Cascading benefits should also be considered as part of the total cost of ownership analysis. For example, does the wash feature or the addition of high-efficiency filters, UV treatments, and pollution control lessen the grease accumulation on the fire system components, ductwork systems, fans, and rooftops, reducing the labor hours for maintenance and cleaning while potentially increasing their life span? This may be especially applicable for the LCC of heat recovery devices (see chapter 4, "Design Considerations"). For additional LCC savings on HVAC equipment, consider higher-efficiency units and variable-speed fans. Other components such as heat exchangers and economizers can reclaim some of the existing energy to help provide a more comfortable working environment year round. With the addition of energy-saving components, automation devices are available to evaluate the building systems as a whole to ensure they are working together. The use of transfer air present in the space for pressurization or for occupancy requirements can also be used for ventilation or mixing instead of adding additional dedicated makeup air fans. With added features, the need for start-up and performance verification are of greater importance. Equal in importance are the demonstrations showing end users the features and functions of these items so that, once installed, the components are used as they were designed and take advantage of the energy savings.

Life-Cycle Component Evaluation

For each system, the total LCC can be calculated by using Equation 1.1 (Fuller and Peterson 1996):

$$LCC = I + Repl - Res + E + W + OM\&R \tag{1.1}$$

where

LCC	=	total LCC in present-value dollars of a given alternative
I	=	present-value investment costs
Repl	=	present-value capital replacement costs
Res	=	present-value residual value (resale value, scrap value, etc.) less disposal costs
E	=	present-value energy costs
W	=	present-value water costs
OM&R	=	present-value non-fuel operating, maintenance, and repair (OM&R) costs

This formula adds the present value of each component and considers escalating utility costs, labor costs, repair costs, etc., in the calculation.

For CKV systems, the lifetime is typically longer than 10 years, with some components lasting the lifetime of the building. If 10 years is used as the span of the analysis, the replacement costs and residual value costs are zero (ASHRAE 2019). Longer time span evaluations may require replacement costs of working items (e.g., fans). The manufacturer can assist with those cost estimates and expected life spans. Other cost considerations could range from fan energy and HVAC costs for conditioning the supply air to water and sewage costs to costs for cleaning and replacing system filters and for ductwork cleaning. The sum of all LCC component costs is the real cost of ownership for the end user.

Life-Cycle Cost Analysis Component Evaluation Example

Table 1.1 shows example costs of three 30 ft (9.1 m) CKV hoods designed for the Dallas/Ft. Worth area with two standard dry hoods and one with an added automatic wash function. The different hoods are evaluated with differing first costs and 10-year E, W, and OM&R costs. For the 10-year evaluation, replacement and residual costs are $0. The LCC calculation shown in Table 1.2 reflects the 30 ft (9.1 m) hood example. The hood with the wash feature, although more expensive initially, has a lower total cost of ownership.

From Tables 1.1 and 1.2, note that the initial costs can vary from differing manufacturers or from a different model from the same manufacturer. Operational calculations take into consideration the water, sewer, and power usage along with hood length, number of filters, and hours and days of operation for both wash and dry hoods. The primary difference in the operational cost comparison is the additional labor hours required to manually remove, wash, or soak the filters, then replace them in the hood on a regular basis.

Each item being evaluated within the HVAC system must not only go through this process but also be applied to the system as a whole. Please note that these calculations may vary in differing parts of the country, even with similar equipment selections.

Load and energy modeling platforms and other tools are commercially available to aid the engineer in calculating the load and system performance from a comfort perspective as well as providing energy and life-cycle costs.

Building System Example

Often kitchen exhaust hoods and exhaust airflows, as well as their design and operating parameters, play an important role in the sizing and layout of the HVAC system. This building system example compares standard-efficiency versus high-efficiency hoods and standard-efficiency versus

Table 1.1 Hood Length, Feature, First Cost, and Operating Cost Operating at Medium Duty[1]

Hood	Length	Feature	First Cost	Yearly Operating Cost[2]
A	30	Dry	$26,000	$15,082
B	30	Dry	$35,000	$15,082
C	30	Wash	$55,000	$5,795

[1]Medium duty as described in ANSI/ASHRAE Standard 154 (2022d)
[2]Restaurant operating 16 h/day, 7 days/week

Table 1.2 LCC Calculation: LCC = I + REPL – RES + E + W + OM&R

Hood	I	Repl	Res	E + W + OM&R 10-Year Evaluation	LCC 10 Years
A	$26,000	0	0	$150,820	$176,820
B	$35,000	0	0	$150,820	$185,820
C	$55,000	0	0	$57,950	$112,950

high-efficiency RTUs on a typical quick-service restaurant (QSR) located in St. Louis, MO. In this example, we used a commercially available load and energy software platform, Trace 3D Plus (Trane 2021), to evaluate the life-cycle costs of several alternative designs. Geography, load variations, and time-of-day loads have a large impact on the building load and energy performance. Reducing exhaust airflows in the high-efficiency hood scenarios reduces the amount of replacement air required and plays a large role in HVAC unit sizing and energy use.

Restaurant type:	QSR
Building layout:	2515 ft^2 (234 m^2)
Location:	St. Louis, MO
Weather data:	ASHRAE 0.4%
Standard-efficiency hood exhaust:	3000 cfm (1416 l/s), exhaust only
	Canopy hood (300 cfm/ft [464.5 l/s per m])
High-efficiency hood exhaust:	2250 cfm (1062 l/s), exhaust only
	Engineered canopy hood (225 cfm/ft [348 l/s per m]) with side curtains
Standard-efficiency HVAC unit sizing:	25 tons (88kW), total (determined by load analysis)
	15 tons (53 kW), kitchen unit (11.0 EER/12.2 IEER)
	10 tons (35 kW), dining unit (11.2 EER/12.7 IEER)
High-efficiency HVAC unit sizing:	22.5 tons (79 kW), total (determined by downsizing load analysis)
	12.5 tons (44 kW), kitchen unit (12.1 EER/14.0 IEER)
	10 tons (35 kW), dining unit (12.1 EER/14.7 IEER)

Figures 1.1 and 1.2 depict the layout of a typical QSR, which was used in the load and energy modeling simulation, and Figure 1.3 indicates the HVAC zoning. There are two zones, Dining and Kitchen, and each is served by a separate RTU.

This example looks at three alternatives as compared to the baseline design, which is detailed in Table 1.3. Figure 1.4 shows the air balance schedule and HVAC unit sizing for each alternative. OM&R cost estimates were the same for all alternatives, as shown in Figure 1.5.

Figures 1.6 and 1.7 show the energy use of the three alternatives versus the baseline design (standard-efficiency exhaust hoods and standard-efficiency, constant-volume HVAC units). Figure 1.8 shows the economic and LCC analysis metrics as compared to the baseline design. For each system, the total LCC can be calculated using Equation 1.1.

Table 1.4 uses the LCC calculation referenced in the "Life-Cycle Component Evaluation" section to depict the LCC in dollars for the baseline and three alternatives modeled. Simple payback, LCC, and internal rate of return (IRR) are shown in Figure 1.8.

Alternative 3 (high-efficiency hoods/high-efficiency variable volume airflow rooftop units) has the lowest LCC of all the alternatives compared to the baseline alternative. However, depending on the first and installed cost sensitivity and the owner's needs, the designer must weigh first cost to LCC to determine the "best fit" system based on the owner's drivers and project constraints.

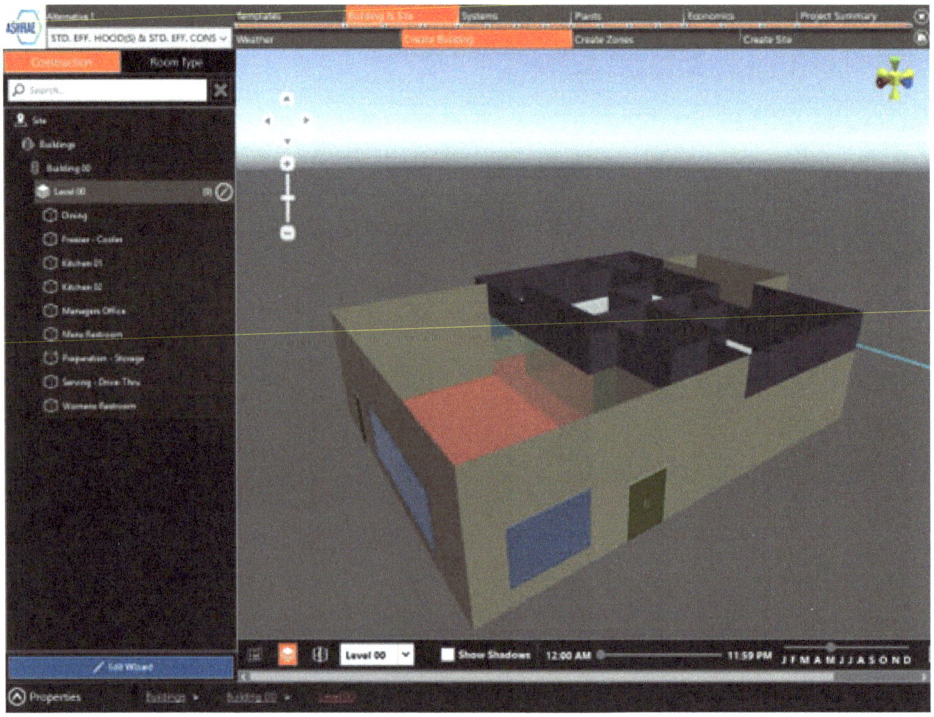

Figure 1.1 QSR restaurant layout produced by Trace 3D Plus (Trane 2021).

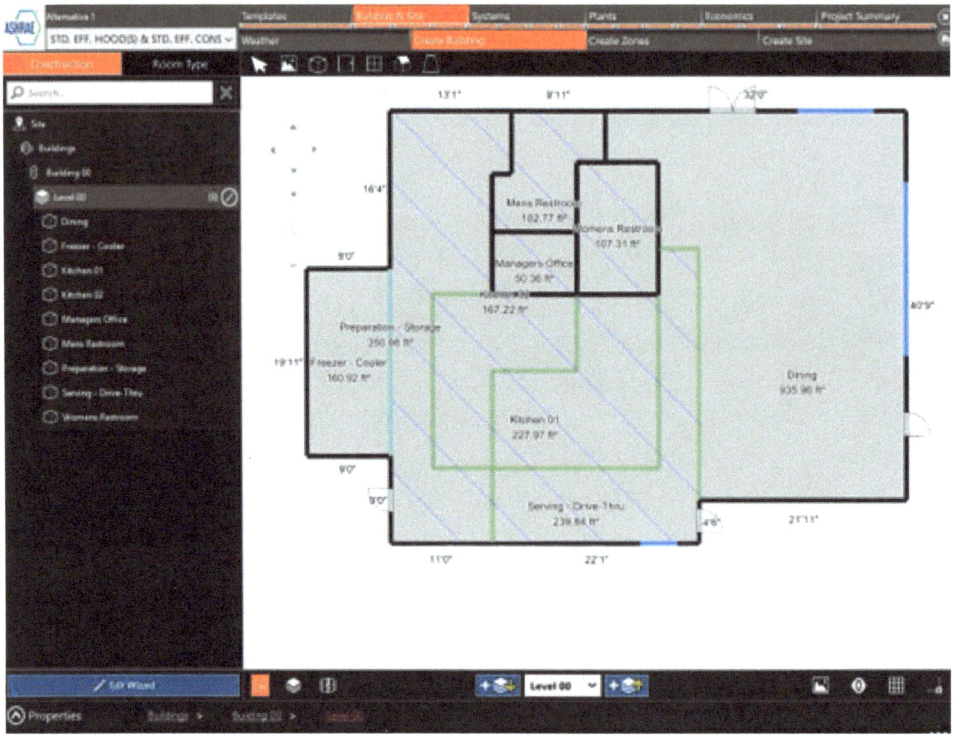

Figure 1.2 QSR restaurant HVAC plan produced by Trace 3D Plus (Trane 2021).

Figure 1.3 QSR restaurant HVAC zoning produced by Trace 3D Plus (Trane 2021).

Table 1.3 Alternatives and First Cost and Installed Cost Input Parameters

System Option	Kitchen Exhaust Hood System	HVAC System	First Cost Equipment and Installation
Baseline	Standard efficiency	Standard efficiency Constant-volume supply air 25.0 tons (88.9 kW)	$67,625
1	High efficiency	Standard efficiency Constant-volume supply air 22.5 tons (79.1 kW)	$71,925
2	Standard efficiency	High efficiency Variable-volume supply air 25.0 tons (88.9 kW)	$74,500
3	High efficiency	High efficiency Variable-volume supply air 22.5 tons (79.1 kW)	$78,800

BASELINE: STANDARD HOOD & STANDARD EFFICIENCY HVAC - AIR BALANCE SCHEDULE

TAG	Supply Air	Return Air	Outside Air	Exhaust Air	Resulting Pressure	% O/Air	Serving	Capacity (Tons)	CFM / Ton
RTU-1	6,000	3,900	2,100	0	2,100	35%	Kitchen	15.0	400
RTU-2	4,000	2,600	1,400	0	1,400	35%	Dining	10.0	400
EF-1	0	0	0	3,000	-3,000	N/A	Griddle	N/A	N/A
EF-2	0	0	0	300	-300	N/A	Restroom	N/A	N/A
TOTAL	10,000	6,500	3,500	3,300	200		CLG. CAPACITY	25.0	
FIRST COST:	$67,625								

ALT. 1: HIGH EFFICIENCY HOOD & STANDARD EFFICIENCY HVAC - AIR BALANCE SCHEDULE

TAG	Supply Air	Return Air	Outside Air	Exhaust Air	Resulting Pressure	% O/Air	Serving	Capacity (Tons)	CFM / Ton
RTU-1	5,000	3,472	1,528	0	1,528	31%	Kitchen	12.5	400
RTU-2	4,000	2,778	1,222	0	1,222	31%	Dining	10.0	400
EF-1	0	0	0	2,250	-2,250	N/A	Griddle	N/A	N/A
EF-2	0	0	0	300	-300	N/A	Restroom	N/A	N/A
TOTAL	9,000	6,250	2,750	2,550	200		CLG. CAPACITY	22.5	
FIRST COST:	$71,925								

ALT. 2: STANDARD HOOD & HIGH EFFICIENCY VARIABLE AIRFLOW HVAC - AIR BALANCE SCHEDULE

TAG	Supply Air	Return Air	Outside Air	Exhaust Air	Resulting Pressure	% O/Air	Serving	Capacity (Tons)	CFM / Ton
RTU-1	6,000	3,900	2,100	0	2,100	35%	Kitchen	15.0	400
RTU-2	4,000	2,600	1,400	0	1,400	35%	Dining	10.0	400
EF-1	0	0	0	3,000	-3,000	N/A	Griddle	N/A	N/A
EF-2	0	0	0	300	-300	N/A	Restroom	N/A	N/A
TOTAL	10,000	6,500	3,500	3,300	200		CLG. CAPACITY	25.0	
FIRST COST:	$74,500								

ALT. 3: HIGH EFFICIENCY HOOD & HIGH EFFICIENCY VARIABLE AIRFLOW - AIR BALANCE SCHEDULE

TAG	Supply Air	Return Air	Outside Air	Exhaust Air	Resulting Pressure	% O/Air	Serving	Capacity (Tons)	CFM / Ton
RTU-1	5,000	3,472	1,528	0	1,528	31%	Kitchen	12.5	400
RTU-2	4,000	2,778	1,222	0	1,222	31%	Dining	10.0	400
EF-1	0	0	0	2,250	-2,250	N/A	Griddle	N/A	N/A
EF-2	0	0	0	300	-300	N/A	Restroom	N/A	N/A
TOTAL	9,000	6,250	2,750	2,550	200		CLG. CAPACITY	22.5	

Figure 1.4 Air balance schedules.

	Investment Costs		Operation, Maintenance & Repair Costs		
Year	Description	Cost	Description	Cost	Grand Total
0	SE HOODS - 25 TON SE CV HGRH ECON	$67,625.00			$67,625.00
1			O,M&R 01	$1,250.00	$1,250.00
2			O,M&R 02	$1,250.00	$1,250.00
3			O,M&R 03	$1,250.00	$1,250.00
4			O,M&R 04	$1,250.00	$1,250.00
5	AIR BALAMCE	$3,500.00	O,M&R 05	$1,250.00	$4,750.00
6			O,M&R 06	$1,875.00	$1,875.00
7			O,M&R 07	$1,875.00	$1,875.00
8			O,M&R 08	$1,875.00	$1,875.00
9			O,M&R 09	$1,875.00	$1,875.00
10	AIR BALANCE	$3,500.00	O,M&R 10	$1,875.00	$5,375.00
11			O,M&R 11	$2,500.00	$2,500.00
12			O,M&R 12	$2,500.00	$2,500.00
13			O,M&R 13	$2,500.00	$2,500.00
14			O,M&R 14	$2,500.00	$2,500.00
15			O,M&R 15	$2,500.00	$2,500.00
	Investment Costs: $74,625.00		OM&R Costs: $28,125.00		Total:$102,750.00

Investment costs (hoods and HVAC system)	First cost, equipment, and installation
	Air balance performed; reoccurs every five years
OM&R costs	Annual cost to operate and maintain the complete HVAC system
	Annual OM&R costs typically vary depending on the age of the equipment and the HVAC equipment's warranty period, as depicted in this figure. For instance, the HVAC equipment's warranty period plays a factor in cost during the first five years; after the warranty period, the cost traditionally increases during mid-life and again as the HVAC equipment nears the end of its life and more maintenance is typically required to keep the equipment in operation. The typical life cycle of a traditional RTU is 15 years.

Figure 1.5 Investment and OM&R cost schedule produced by Trace 3D Plus (Trane 2021).

Alternative Comparison

Energy End Use Comparison

		Baseline		Comparison 1		Comparison 2		Comparison 3	
		STD. EFF. HOOD(S) & STD. EFF. CONSTANT VOLUME RTU		HIGH EFF. HOOD(S) & STD. EFF. CONSTANT VOLUME RTU		STD. EFF. HOOD(S) & HIGH EFF. VARIABLE VOLUME RTU		HIGH EFF. HOOD(S) & HIGH EFF. VARIABLE VOLUME RTU	
		Energy (kBTU)	% of Total	Energy (kBTU)	% Improve	Energy (kBTU)	% Improve	Energy (kBTU)	% Improve
Heating	*Natural Gas*	734,596	40%	601,428	18%	483,481	34%	386,216	47%
Cooling	*Electricity*	213,392	12%	191,307	10%	150,134	30%	135,680	36%
Interior Lighting	*Electricity*	48,699	3%	48,699	0%	48,699	0%	48,699	0%
Exterior Lighting	*Electricity*	152,437	8%	152,437	0%	152,437	0%	152,437	0%
Interior Equipment	*Electricity*	325,698	18%	325,698	0%	325,698	0%	325,698	0%
	Natural Gas	257,285	14%	257,285	0%	257,285	0%	257,285	0%
Exterior Equipment	-	-	-	-	-	-	-	-	-
Fans	*Electricity*	118,344	6%	105,625	11%	36,074	70%	32,188	73%
Pumps	-	-	-	-	-	-	-	-	-
Heat Rejection	-	-	-	-	-	-	-	-	-
Humidification	-	-	-	-	-	-	-	-	-
Heat Recovery	-	-	-	-	-	-	-	-	-
Water Systems	-	-	-	-	-	-	-	-	-
Refrigeration	-	-	-	-	-	-	-	-	-
Generators	-	-	-	-	-	-	-	-	-
Summary	*Electricity*	858,571	46%	823,767	4%	713,043	17%	694,702	19%
	Natural Gas	991,881	54%	858,713	13%	740,766	25%	643,501	35%
	Total	1,850,452	100%	1,682,479	9%	1,453,809	21%	1,338,204	28%

Figure 1.6 System alternative energy consumption comparison breakdown produced by Trace 3D Plus.
Courtesy of Trane (2021)

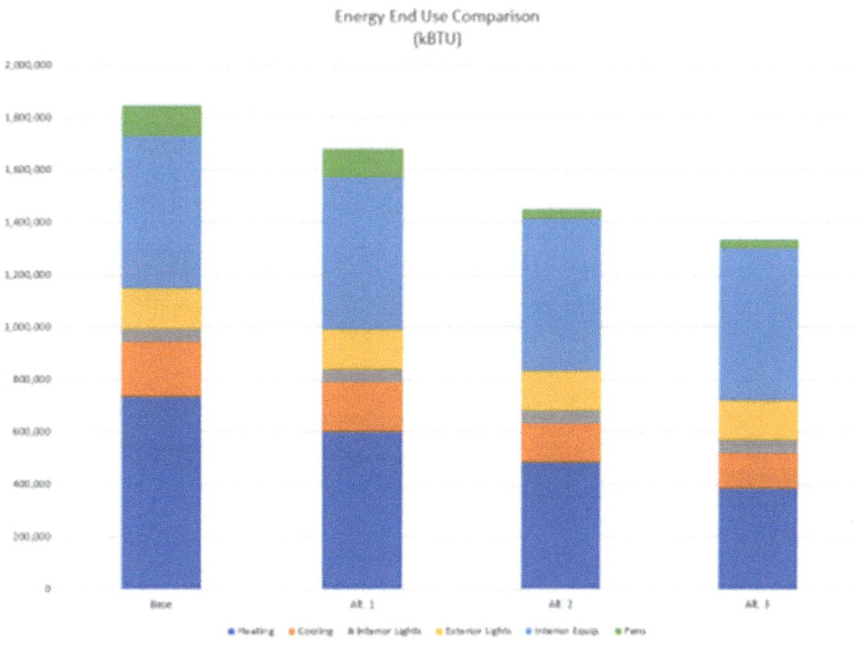

Figure 1.7 Total system alternative energy consumption comparison.

Comparison	Initial Investment Cost Difference ($)	First Year Utility Savings ($)	Simple Payback Period	Life Cycle Payback Period	Internal Rate of Return (IRR)
Comparison 1 vs Baseline	4,300.00	2,453.81	1.75 Years	2.77 Years	56.96%
Comparison 2 vs Baseline	6,875.00	7,203.00	0.95 Years	1.95 Years	104.77%
Comparison 3 vs Baseline	11,175.00	8,766.82	1.27 Years	2.28 Years	78.43%
Comparison 2 vs Comparison 1	2,575.00	4,749.19	0.54 Years	1.54 Years	184.43%
Comparison 3 vs Comparison 1	6,875.00	6,313.01	1.09 Years	2.09 Years	91.82%
Comparison 3 vs Comparison 2	4,300.00	1,563.82	2.75 Years	3.83 Years	35.87%

Figure 1.8 Alternative life-cycle comparison analysis produced by Trace 3D Plus.
Courtesy of Trane (2021)

Table 1.4 Alternative Life-Cycle Cost Calculations

Alternative	LCC	I	Repl	Res	E	W	OM&R
Baseline	$698,790	$74,625	$0	$0	$596,040	$0	$28,125
1	$664,195	$78,925	$0	$0	$557,145	$0	$28,125
2	$591,485	$81,500	$0	$0	$481,860	$0	$28,125
3	$570,690	$85,800	$0	$0	$457,065	$0	$28,125

2

Predesign

Objective

The design of a commercial kitchen ventilation (CKV) system begins with a complete understanding of how the facility is being used. This includes the following:

- Location of kitchen
- Type of kitchen/restaurant
- Types of food being prepared
- Hours of operation
- Occupancy
- Ventilation air
- Exhaust airflows for hoods
- Comfort and safety
- Life-cycle costs
- Codes and standards
- Testing, adjusting, and balancing (TAB) and commissioning

With this knowledge, a designer can begin the process of selecting the equipment and services required for a successful kitchen ventilation system. This understanding will also help the designer understand the owner's requirements and maximize the project's success. A well-designed kitchen ventilation system can lead to years of successful, comfortable, and safe operation for all occupants.

Location of Kitchen

The location of the restaurant plays an important role in how busy it will be and when. For instance, a quick-service restaurant (QSR) near office buildings can be extremely busy for weekday lunch and less busy on a weekend, while the same restaurant in a shopping mall food court might be busiest during lunch on weekends and less busy on weekdays. A restaurant's location is an important variable to predict future business.

Another variable that comes into play when selecting a restaurant's location is its neighbors. This may be brought up during zoning; at other times, it occurs after the restaurant is built. The grease effluent from the restaurant's exhaust system can leave residue on its building or adjacent buildings, and the odors can be a nuisance for nearby residential and/or commercial properties. Addressing the effluent and odor leaving the building should be considered during the predesign phase of a commercial kitchen. In a city environment, a designer may be forced to consider a pollution control unit (PCU) based on local codes that restrict cooking odors and smoke from commercial kitchens.

Type of Kitchen/Restaurant

Understanding the type of restaurant, food types, food preparation, and hours of operation are all important for good CKV design. Each restaurant and restaurant type can vary in any of those characteristics. There is not a one-size-fits-all approach to restaurant CKV design. The following list describes some common kitchen and restaurant types:

- **QSRs.** Commonly referred to as "fast-food restaurants," QSRs are typically the smallest in footprint, and their focus is on speed and convenience. They can consist of dine-in, pick-up, walk-up, or drive-through areas or any combination of those. A majority of QSR business occurs in the drive-through area. QSR business is about consistency and repeatability, so their ventilation design is efficient and often standardized. QSRs are usually chain restaurants.
- **Fast-casual restaurants.** Fast-casual restaurants are similar to QSRs but offer a more upscale experience such as plated meals. Though some may have drive-throughs, most of the business is likely from dine-in customers. Examples of fast-casual restaurants include cafés, buffets, and some chain restaurants.
- **Full-service restaurants.** Full-service restaurants fall into a variety of categories including hospitality, pubs, bars, family, and fine dining. In these restaurants, all service occurs at the table by a server.
- **Ghost kitchens.** Also known as "off-site" or "digital" kitchens, these are restaurant locations with no storefront. They focus solely on deliveries. A ghost kitchen's design is the same as that of drive-through only or carry-out only restaurants. They are an extension of the type of restaurant they serve, and their busiest parts of the day will mirror that.
- **Tenant restaurants.** Tenant restaurants present a unique design challenge. These are restaurants in which some of the cooling and replacement air is dependent upon the landlord (e.g., a restaurant located within an indoor mall food court may rely on the mall to provide transfer air from the common area). For example, when designing a ventilation system for a tenant restaurant, the engineer may be given parameters to the space that limit the supplied conditioned airflow. When a restaurant's ventilation system is being designed in a tenant space, there may not be enough available transfer air or cooling capacity at the start of the design, and the engineer may have to consider adding additional cooling capacity.
- **Special or unique restaurants.** A restaurant's cuisine provides insight into grease generation and hood selection. For instance, wok cooking is one of the largest emitters of grease effluent. Display cooking, common to Japanese hibachi restaurants, requires substantially higher exhaust rates due to the use of island hoods.
- **Food trucks.** Food trucks present serious design and safety challenges. Because they typically rely on open windows for replacement air, wind can negatively impact the hood's ability to properly capture and contain the cooking effluent. Many food trucks also use mini-split systems or packaged terminal air conditioners (PTACs) for cooling at the hood. That airflow directed at the hood also impacts capture and containment (C&C). Food trucks are not covered in this guide.

Types of Food Being Prepared

To properly design a CKV system, the designer must know the types of food being prepared. This is critical information to select the required appliances and determine the airflow of the ventilation hood and any other required HVAC devices.

Hours of Operation

There are four common hours of operation used in food service: breakfast/brunch, lunch, dinner, and late night. In some cases, a restaurant may serve breakfast all day or may be closed to customers during other periods of the day. Knowing how and when a restaurant will operate leads to a better design. It is always best to ask the project's owner what hours of operation the restaurant will have and the details of their operations. Typical examples are listed below.

- **Breakfast/brunch.** Breakfast foods can vary greatly based on the cuisine being served. Breakfast is commonly served from 6:00 a.m. to 11:00 a.m. In some restaurants, breakfast may be an all-day affair. In that scenario, there may be additional appliance heat gains to consider. For instance, a restaurant that uses eggs only for breakfast may have a dedicated egg cooking station that will be ready to use to serve breakfast all day. For QSRs and fast-casual restaurants, breakfast peaks are Monday through Friday, and nearly all business occurs in the drive-through. For full-service restaurants, peak breakfast/brunch is usually Sunday mornings.
- **Lunch.** Lunch service typically occurs from 11:00 a.m. to around 3:00 p.m. For QSRs and fast-casual restaurants, lunch and dinner menus are nearly identical. For full-service restaurants, lunch is likely a different menu; it may contain similar or some of the same items as dinner but in smaller portions. It may also have items that are designed for fast cooking and service focused on the business lunch hour.
- **Dinner.** Dinner service typically occurs from 5:00 p.m. to 10:00 p.m. Some restaurants may only offer dinner but may have preparatory time periods earlier in the day that also require ventilation.
- **Late night.** Late-night service typically occurs after 10:00 p.m. For a QSR, this may be drive-through only, or the dining room may be open. For some pubs or bars, hot-food service may be discontinued during this time. For hospitality, the restaurant may be closed but the kitchen may stay open for room service.
- **Hours for ancillary spaces.** Another important design factor concerning the hours of operation is how ancillary spaces will be handled. Other spaces may have a separate HVAC system and operating schedule.

Occupancy

Designers should be aware of when a commercial kitchen will be in operation, determined by the type of facility and meals served. Occupancy profiles for the kitchen and dining room are considerably different and should be considered separately. Restaurant staff arrive earlier and leave later than the restaurant's hours of business. For example, a QSR is open from 6:00 a.m. to 11:00 p.m. The staff arrive about an hour prior to opening and leave about an hour after closing. Staffing hours vary, with a combination of full- and part-time employees based on the restaurant's expected business load for the day. The weekday lunch hour may be the restaurant's busiest time, followed by breakfast and then dinner. To design a CKV system, the designer should know the hours of operation to determine the schedule for all the HVAC systems in the facility. Figure 2.1 shows an

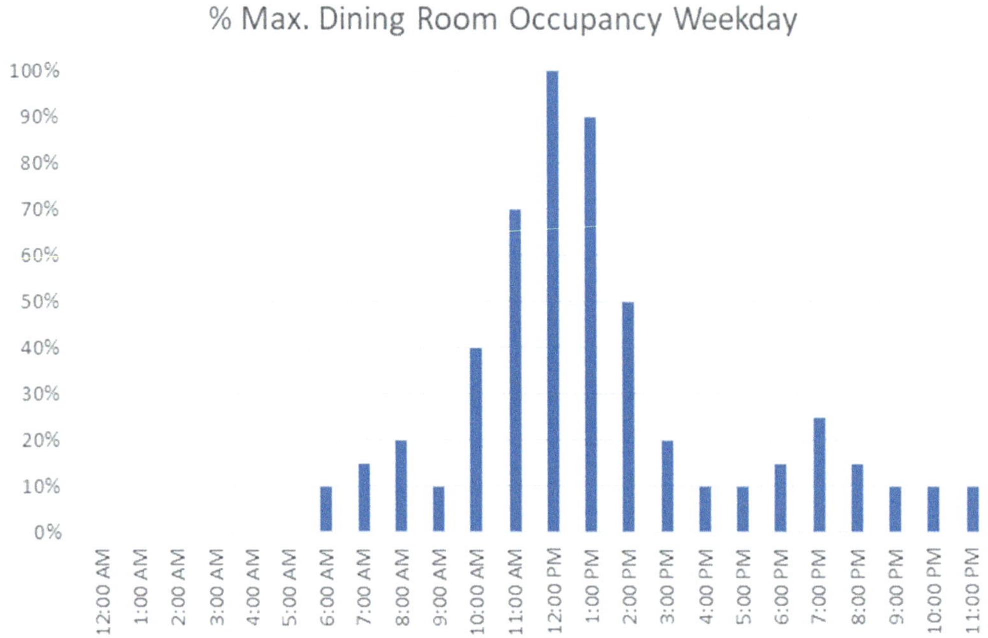

Figure 2.1 Example dining room occupancy profile for a QSR (weekdays).

example occupancy schedule for a freestanding restaurant only open during weekdays. This information should be used to schedule the HVAC devices.

Ventilation Air

It is important that a designer consider the pressurization in commercial kitchen facilities. The HVAC design should ensure the kitchen is positive pressure to the building's exterior but negative pressure relative to its adjacent interior spaces to prevent the migration of odors. Some codes may dictate the maximum pressure differential between spaces. Additionally, food service areas in healthcare facilities may have their own requirements. Chapters 3 and 4 discuss designing a kitchen ventilation system to maintain proper ventilation air and pressures.

Exhaust Airflows for Hoods

Having properly sized exhaust airflow for kitchen hoods is extremely important. The exhaust through a hood removes the contaminants that are generated from the cooking process. The exhaust into a hood also removes contaminants that can be generated in the kitchen itself, thus improving the comfort and safety of the restaurant staff. Properly sized exhaust also ensures that odors and cooking byproducts are kept from entering the dining area where customers are eating. This guide discusses proper hood exhaust in chapters 3 and 4.

Comfort and Safety

Controlling the temperature and humidity of a commercial kitchen is critical for both kitchen workers and servers. The resulting space temperature and ventilation can affect workers' and servers' productivity and health. Chapter 4 addresses the design of a system to obtain comfort, and chapter 5 discusses health and safety issues in a commercial kitchen.

Life-Cycle Costs

The costs associated with a CKV system include not only the first cost for the equipment and labor for installation but also the costs involved throughout the life of the system. Life-cycle costs are discussed in detail in chapter 1.

Codes and Standards

A CKV system comprises a number of components. The equipment, including ventilation hoods, ductwork, fire suppression systems, controls, and other components, all have codes and standards that are covered within this guide.

TAB and Commissioning

TAB and commissioning are verifications that the CKV system's functionality and design have been properly implemented. See chapter 8 for information on TAB and commissioning.

3

Exhaust Hoods and Related Equipment

The foundation of any sustainable food service cooking operation is a properly engineered commercial kitchen ventilation (CKV) system. The kitchen's exhaust fan(s), Type I/Type II hoods, fire protection system, and supporting replacement air systems govern the facility's comfort and safety as well as nearly 30% of its energy consumption (DOE 2015). The kitchen's hoods are the heart of that system, with their successful specification ultimately scaling the whole CKV system's infrastructure and size. Whether designing for versatility, performance, location, or space, all factors are considered in the available hood options in this chapter. This chapter covers what you need to know about Type I and Type II hoods to properly optimize energy, comfort, performance, operations, maintenance, and repair and provide a strong foundation for building a high-performance CKV system.

Defining Hood Loading

The process of designing commercial kitchen hoods starts with the cooking equipment placed beneath them. Based on the type of emissions (grease or steam) and the intensity of those emissions, a duty can be assigned to each piece of cooking equipment. That duty level becomes the foundation for setting the type, style, and ventilation rates for the hoods. As the equipment sets the requirements for the hoods, the hoods then set the requirements for the other CKV components, including the fire protection system, exhaust ductwork, exhaust fan(s), and replacement air systems.

For example, if a designer exchanges a medium-duty grooved griddle for a heavy-duty gas broiler under a 5 ft (1.52 m) hood to cook hamburger, the substitution would increase the exhaust airflow by 33%, which would require a 33% larger exhaust duct system, a new fan to move 33% more air at ~33%+ higher exhaust motor horsepower (kilowatts), and an increased capacity in the replacement air system tempering/conditioning to accommodate the added 33% more replacement air. Depending on the local climate, some facilities would easily see greater than one ton of added conditioning capacity to accommodate the additional replacement airflow and added heat gain emitted to the space due to the radiant heat output of the gas broiler. (To better understand the impact of space heat gain and CKV heating and cooling load calculations, refer to chapter 18, "Nonresidential Cooling and Heating Load Calculations," of *ASHRAE Handbook—Fundamentals* [ASHRAE 2021b]).

It should be noted that the categorization of cooking equipment has been studied rigorously for over two decades, resulting in values by appliance description for grease produced in pounds of emissions per 1000 lbs. cooked (ASHRAE 1998), radiant heat gain to space (ASHRAE 2008), and duty classifications regarding exhaust ventilation rates. Reference sources for this data can be found in the following publications:

- Chapter 18, "Nonresidential Cooling and Heating Load Calcualtions," *ASHRAE Handbook—Fundamentals* (2021b)
- Chapter 34, "Kitchen Ventilation," *ASHRAE Handbook—HVAC Applications* (2019)
- ANSI/ASHRAE Standard 154 (2022d), Type I and II duty classifications

Grease Producing Equipment and Type I Hoods

Understanding the effluent and grease produced during cooking is critical when selecting the exhaust hood(s). NFPA 96 defines grease as "rendered animal fat, vegetable shortening, and other such oily matter used for the purposes of and resulting from cooking and/or preparing foods" (2021b, 96-10). Grease is more specifically characterized in ASHRAE Research Project (RP) 745 (Gerstler et al. 1998). The presence of grease emissions produced from a cooking process will dictate the need for a Type I hood and exhaust system. Grease is a known fire liability due to its high potential energy content when spread throughout a commercial kitchen exhaust system. The requirement of a Type I hood can be avoided altogether when the production of grease from the appliance is tested to levels at or below 3.1×10^{-7} lb/ft^3 at 500 cfm (5.0 mg/m^3 at 236 L/s) when tested to UL 710B (2011a) requirements and the generated heat gains are accounted for in the HVAC system.

Non-Grease Producing Equipment and Type II Hoods

Equipment with primary emissions of steam or convective heat can be served by a Type II hood system. Equipment identified in ASHRAE Standard 154 (2022d) as "Hood Not Required" are acceptable for installation without a hood when the emissions are below 3.1×10^{-7} lb/ft^3 (5.0 mg/m^3) at 500 cfm (236 L/s) when tested to UL 710B requirements and the sensible and latent loads of the equipment are accounted for in the HVAC design capacities. Refer to chapter 18, "Heat Gain," in *ASHRAE Handbook—Fundamentals* (2021b).

Examples of equipment not requiring a hood are espresso machines, electric panini presses, under-counter dishwashers, electric hot dog cookers, popcorn cookers, and more. Examples of equipment requiring a Type II hood with a "Light Duty" classification are steamers, rice cookers, and kettles under 20 gal capacity. Equipment requiring Type II hoods with a "Medium Duty" classification are conveyor dishwashers, gas pasta cookers, and steam-jacketed kettles over 20 gal capacity. See Table 2 in ASHRAE Standard 154 (2022d) for more information on Type II hood requirements for specific appliances.

Determining Hood Duty Level

See Figure 3.1 for a flowchart on determining hood type and hood duty level.

Type II Hoods

ASHRAE Standard 154 (2022d) describes a Type II hood as one that collects and removes steam, heat, and the products of combustion where grease or smoke is not present. Also known as "vapor" or "condensate" hoods, these hoods are designed primarily for the capture and removal of condensate from steam cooking, dish washing, and baking operations where appliance emissions are free from grease. Hoods of this type are referred to as *heat* or *fume* hoods. When conditions allow for the use of a Type II hood, they can be an attractive option in the design process due to the sometimes-significant reduction in system cost over Type I hoods, which require grease- and fire-rated filters, hoods, exhaust ductwork, exhaust fans, and fire suppression.

Some common constructions for Type II hoods are shown in Figures 3.2 and 3.3. Such hoods are typically designed to capture and drain off condensed water to floor drains or condensate gut-

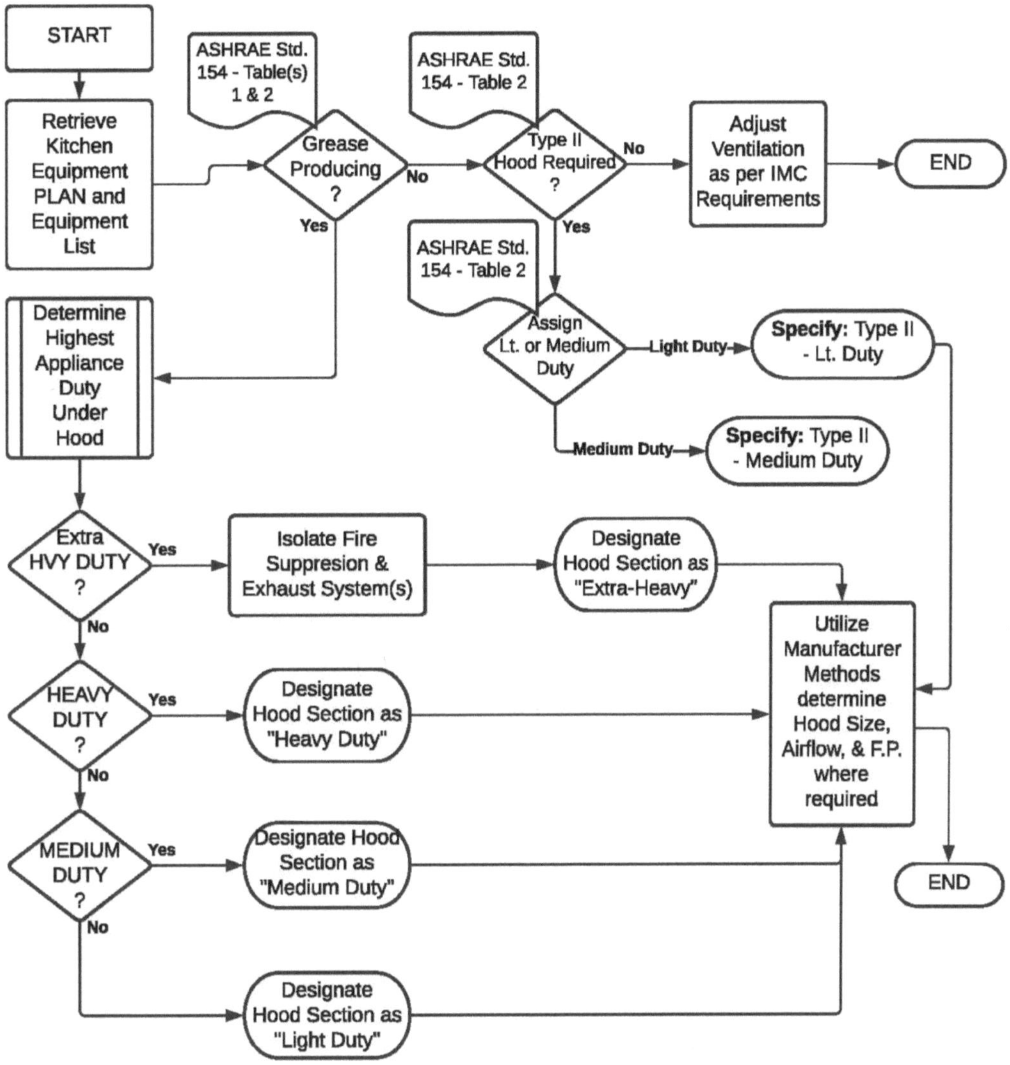

Figure 3.1 Determining hood type and duty.
Courtesy of Gaylord Industries

ters on the hood's perimeter where condensate is evaporated over time back into the hood's exhaust.

Type II Hoods—Current Design Processes and Code Requirements

Type II hoods are not covered under any product-listing category nor do they carry a fire rating. The Type II hood design process focuses primarily on hood overhangs and designing adequate ventilation rates. Note that the minimum airflow for a dishwasher shown in the code may not be high enough for the dishwasher type being used. Consult the hood manufacturer for design guidance. Applied airflow rates are in cubic feet per minute per linear foot of hood (cfm/linear ft) or liters per second per linear meter of hood (L/s per linear m). The process for selecting the total hood section ventilation rate is determined by a simple equation:

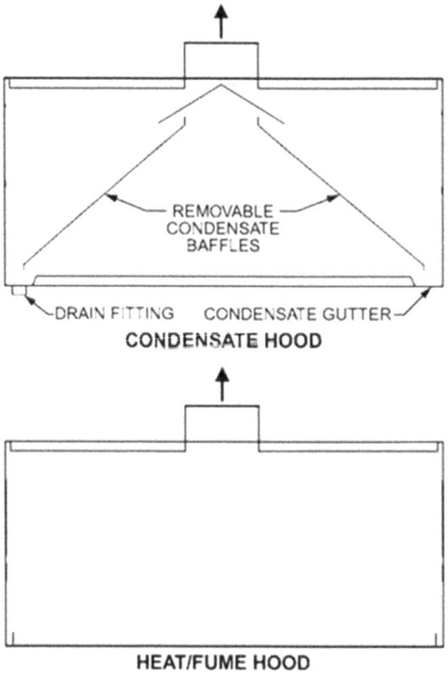

Figure 3.2 Condensate hood construction.

Courtesy of ASHRAE (2019)

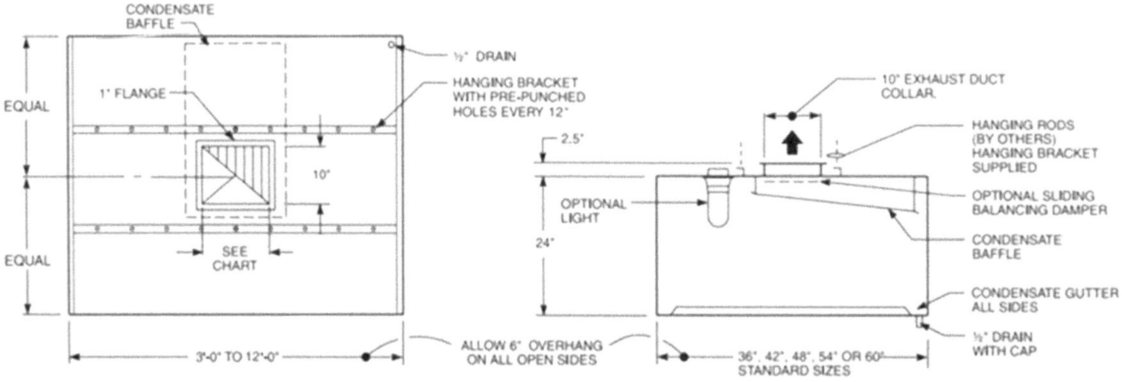

Figure 3.3 Condensate hood example.

Courtesy of Gaylord Industries

$$\text{cfm} = \text{cfm/linear ft of hood} \times \text{hood length in ft}$$

or

$$\text{L/s} = \text{L/s per linear m} \times \text{linear hood length in m}$$

Note that the cfm/linear ft guideline is set by the governing mechanical code based on the equipment covered. Some older standards may apply the airflow based on an area value to set the exhaust rate.

Meeting code minimum airflows on a Type II hood is set by the model mechanical codes, with no current exemption for "listed" airflows or sizing as is available with Type I hoods. Mechanical codes do document the required minimum hood overhangs, maximum hood height above appliance, general material/construction requirements, and generic minimum exhaust airflow rates.

For example, the *International Mechanical Code* (IMC; ICC 2021d) requires a wall hood be sized at no fewer than 6 in. (152 mm) horizontally beyond the top surface of the appliance on the front and ends. It also requires a minimum airflow rate of 100 cfm/linear ft (155 L/s per linear m) across the length of the hood. As previously stated, appliances not requiring a hood, as identified in the mechanical code, require that their heat gains be accounted for in the HVAC loads for the space and that the space exhaust rate be increased. All designs must be confirmed as meeting or exceeding the local code requirements to achieve permitting for the project.

Type II Hoods—Meeting NSF Requirements

Where food is being prepared under a hood, it must also be in compliance with NSF/ANSI 2 (NSF 2021), which requires materials and construction methodology promoting cleanable, antimicrobial surfaces and joints, especially in those areas where condensate might drip back onto the food being prepared beneath.

Type II Hoods—Design Considerations

While Type II hood design does not incorporate the complexity present in Type I (grease producing) applications, it does have opportunities for thoughtful design to address what can sometimes be extremely large latent loads introduced into the facilities. ASHRAE RP-1778 (Livchak and Swierczyna 2020) reports that research determined that applying current model code design processes to the design of a hood can result in up to 50% of a dish machine's latent and sensible outputs escaping into the space. The research determined that increasing hood airflow rates and overhangs has strong impacts on capturing those escaping loads. It also identified typically overlooked sources of heat and moisture that escape into the space, such as dish sanitizing and prewash processes, all of which must be factored into the HVAC design process. The process outlined in Figure 3.4 will remove over 85% of dish machine and fume equipment loading in the space, leaving only 15% to be incorporated into the building's HVAC design. Refer to chapter 18, "Nonresidential Cooling and Heat Load Calculations," in *ASHRAE Handbook—Fundamentals* (2021b) for a list of common appliance heat gains. Confirm loading through ASHRAE research or equipment manufacturer data.

Figure 3.4 shows a method to exceed the current code performance shortfalls and reduce heat and moisture buildup in a facility. The process applies the performance enhancements and ventilation rates resulting from ASHRAE RP-1778 (Livchak and Swierczyna 2020) to achieve a more sustainable performance improvement to the comfort of facilities.

Type II Hoods—Commissioning

Prior to turning over the facility to the kitchen's new owner, the performance of both Type I and Type II hoods is required to be field verified under ASHRAE Standard 154 (2022d). Using the manufacturer's methods, each hood's design airflow is to be set and confirmed, followed by a visual confirmation of each hood's successful capture and containment (C&C) of all steam, convective heat, and products of combustion. This process is to be accomplished over fully heated cooking surfaces under operational conditions or through a complete cycle for dish machines. All sources of replacement air are to be active throughout the process and operating inside of –0.02 to 0 in. of water (5.0 Pa). Refer to chapter 8, "TAB and System Commissioning," for details.

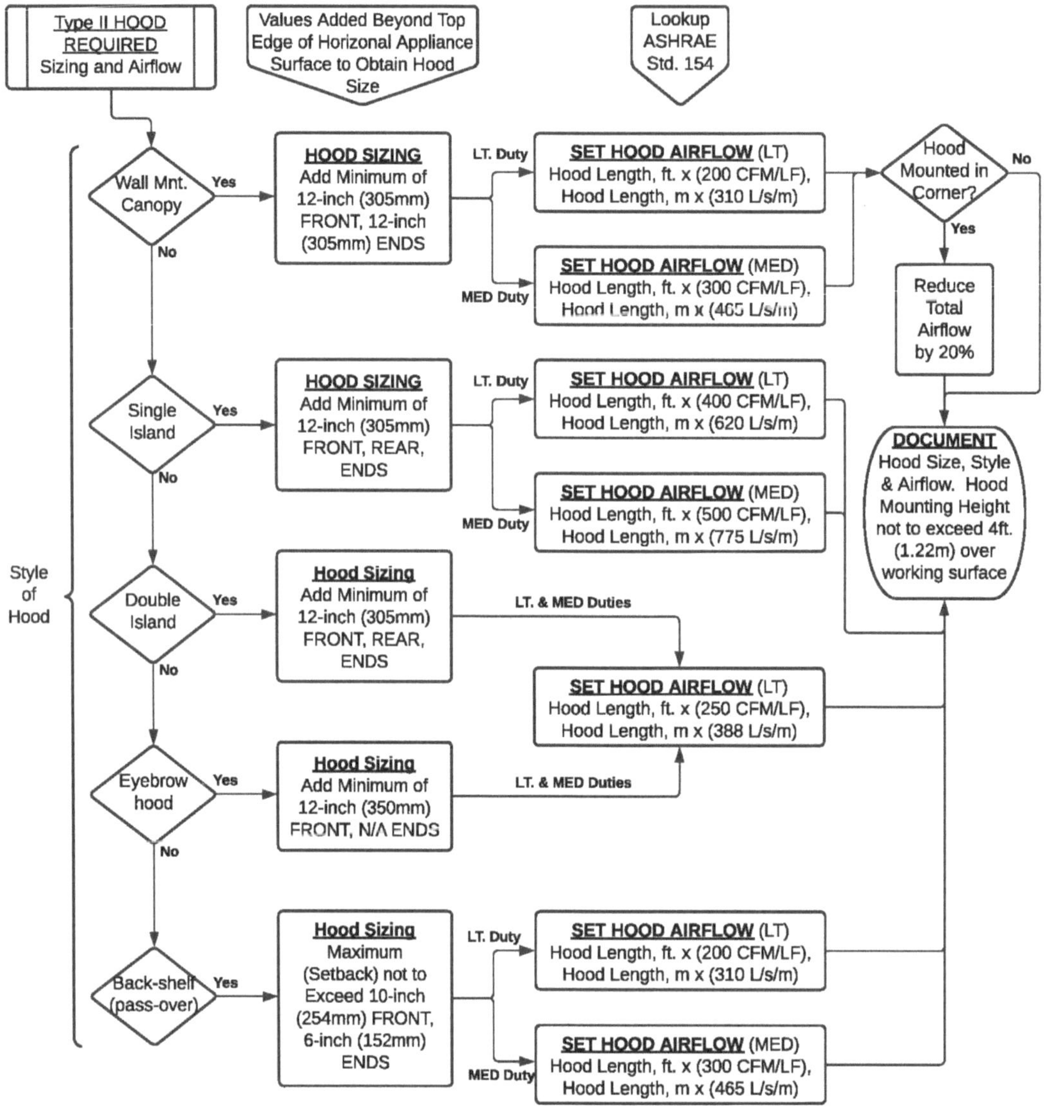

Figure 3.4 Type II design process flowchart.
Courtesy of Gaylord Industries

Type I Hoods

When grease-producing cooking equipment is installed, a Type I hood is required. While Type I hoods require attention to overhangs, mounting heights, and equipment-duty-based airflows like Type II (vapor/condensate) hoods, they also require listed grease filters, listed fire suppression systems covering equipment surfaces, ducts, plenums, fire-rated downstream ductwork, and safe clearances to combustibles. The hoods themselves are listed, allowing for the specific application of airflows, mounting, overhangs, clearance to combustible materials, fire dampers, makeup air compensation, and grease mitigation strategies such as water wash, ultraviolet-C (UVC), electrostatic precipitators, water scrubbers, and more. Type I hoods are supported by fire-rated or listed Type I ductwork, listed canopy lighting systems, listed commercial kitchen exhaust fans, listed commercial

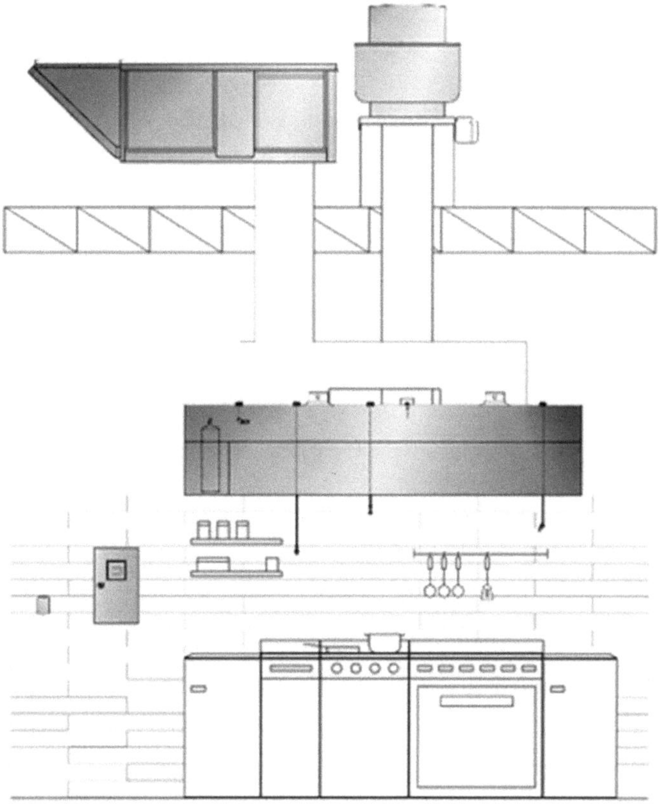

Figure 3.5 Typical interconnected system for Type I hoods.
Courtesy of Accurex, a Greenheck Group Company

kitchen demand controls, and, where locally required, listed commercial kitchen pollution control units (PCUs) based on recognized filters, electrostatic precipitators, scrubbers, incinerators, and more. (For additional information on these associated listed items, see chapter 4, "Design Considerations"). Figure 3.5 highlights how primary components of these systems are interconnected.

Type I hoods have several discrete features, all of which have separate performance, safety, and in some cases, sanitation ratings associated with them. They typically include a canopy, a sloping plenum, and an exhaust collar with accessories such as grease filters, a grease cup, optional fire dampers, and canopy lights. Type I hoods incorporate mounting points to hang the hoods from the ceiling and commonly insulated air spaces on the hood's perimeter to allow for mounting to combustible or limited combustible surfaces inside the kitchen. Generally, Type I hoods are built in sections ranging from 3–16 ft (0.91–4.88 m). Longer hoods typically line up multiple hoods end to end to generate a larger hood system with multiple duct collars (grease duct attachments) to draw the exhaust air through the hood and out of the space. Figure 3.6 shows basic Type I hood components.

A starting place for the classification and construction of Type I hoods is found in NFPA 96 (2021b), which allows for two paths to compliance: listed or unlisted.

- **Listed hoods.** The first path forward allowed under NFPA 96 (2021b), though not an option under ASHRAE Standard 154, is to provide a listed hood. For the purposes of this design guide, *listed* means the provided equipment is publicly identified as following the construction, materials, or performance standards specified in a consensus standard specific to the application where it is being used by an agency acceptable to the authority having jurisdiction (AHJ), who conducts regular audits of the provided products or components. The product listing specific to Type I hoods is UL 710 (2012) or, where UV systems are applied, UL 710C (2006). Hoods known as *ventless* or *recirculating* are covered under UL 710B (2011a). This standard establishes alternative canopy sizing, joints, grease removal devices, dampers, airflows, and more that are equivalent or better in safety and performance when compared to an unlisted hood. The proven construction, installation, and airflow rates are then used as requirements for manufacturers to design, construct, and provide commercial kitchens wherever the model mechanical codes are applied. Listed products by their very nature complicate a generic design process due to variations in manufacturer product performance, sizing, ratings, and more.

It is strongly recommended that manufacturers by used for the application engineering of listed Type I hood systems. They are the best sources to provide optimal, code-compliant solutions and accountability for those designs. The position of ASHRAE Technical Committee (TC) 5.10 is that it is best to allow only listed hoods, as they have been third-party tested and verified to provide lower air and proven construction methods.

- **Unlisted hoods.** The second path forward under NFPA 96 (2021b) is to provide an unlisted hood. An unlisted hood has specific requirements for steel of not less than 18 MSG, 0.048 in. (1.22 mm), or stainless steel of not less than 20 MSG, 0.036 in. (0.91mm), to be continuously welded, greasetight and watertight to the lower, outermost perimeter of the hood. Such hoods require listed grease filters and listed canopy light fixtures and do not allow for the application of fire or balancing dampers. The position of the equipment's cooking surface to the lowermost point of the hood filter opening is not be to less than 18 in. (457 mm) for light- to heavy-duty cooking appliances and not less than 4 ft (1.22 m) when covering cooking equipment utilizing solid fuel for all or part of its heat source. They require listed fire protection and are to be sized and configured to provide for the capture and removal of grease-laden

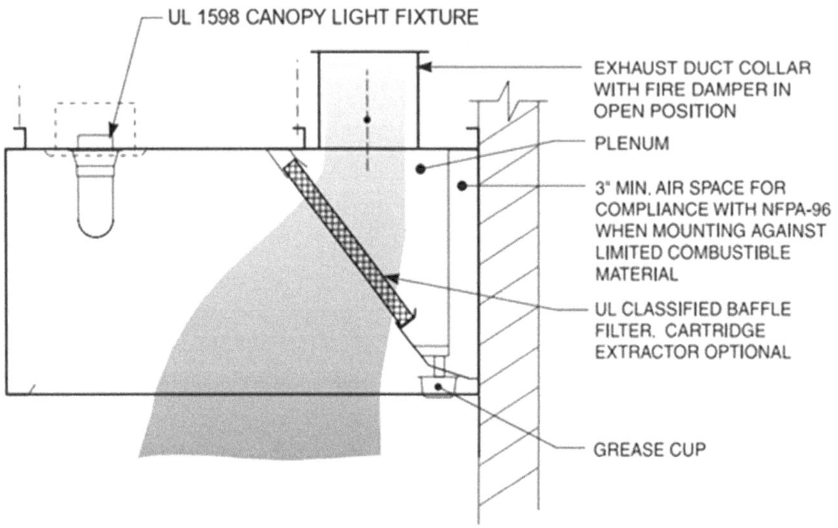

Figure 3.6 Basic Type I hood components.
Courtesy of Gaylord Industries

vapors. The use of an unlisted hood is not accepted under ASHRAE Standard 154 (2022d), as it is proven to use greater exhaust flow rates than a listed hood.

Type I Hoods—Components and Subassemblies

Commercial kitchen hoods rely on specific listed, recognized, or classified components and subassemblies to function safely and properly inside the CKV system. Below are several commonly provided components and subassemblies specified for commercial kitchen Type I hoods.

Fire suppression. Due to the nature of commercial cooking equipment and their associated emissions' potential to burn markedly when conditions are right, NFPA 96 (2021b) requires the application of a fire suppression system to a Type I hood system. These assemblies are typically manifolded wet chemical or water-based systems plumbed through the hoods to deliver a suppressant to the surface of the cooking equipment, inside the plenum of the hoods, and to the hood duct collar transition into the exhaust duct. Downstream coverage of the grease duct varies by manufacturer and is typically up to 75 ft (22.9 m), at which point additional nozzles may be required for the downstream ducting. The governing standard for listed commercial kitchen fire suppression is UL 300 (2019), which is covered under NFPA 17A (2021a) and required by NFPA 96 (2021b). Listed water-based systems are also available under UL 300 (2019), UL 199 (2022a), or UL 199E (2017a). Units covered under UL 199 and 199E are integrated into the building's fire sprinkler system, covered under NFPA 13 (2022).

The design of fire suppression is to be conducted by a properly trained and certified agent of the equipment manufacturer and performed in accordance with the product's listing. Hood manufacturers and fire protection contractors are commonly good sources for these system designs. System sizing for duct and plenum coverage typically scales with hood length and exhaust duct collar sizing, while equipment surface protection scales based on the type and quantity of equipment pieces covered. There are two common types of coverage methods: equipment specific and flood. Equipment-specific coverage is detailed based on a known cooking line requiring modification if equipment is repositioned, added, or deleted from the line. Flood coverage allows for a generic cover over most equipment, but at a higher upfront cost to the project. Depending on which path is taken, surface protection is specifically detailed in product listing manuals and must be closely followed. The cooking lineups and methodology determine the system's cost and sizing.

Special attention must be provided for large kitchens supporting multiple hoods installed end to end or back to back on manifolded Type I duct systems. It is not uncommon for the design of these systems to be actuated simultaneously due to their being in fluid connection with one another, a condition referred to as a common or single hazard. Downstream, pollution control, heat recovery, or other grease duct-mounted accessories are included in this single hazard and must be confirmed and designed in accordance with the locally adopted code. Pollution control or other duct accessories are required to have UL 300 or equivalent protection as required under NFPA 96 (2021b). When the coverages of these devices are not detailed specifically in UL 300 (2019) or an equivalent listing, special attention and AHJ approval is required prior to permitting and installation.

An example system might have UL 300 fire protection equipment provided with tanks, detectors, and nozzles for equipment surface, plenum, and grease duct discharge. Tanks can be optionally hood mounted, as shown in Figure 3.7. Once a fire is detected, the chemical will release simultaneously across the entirety of the common hazard, thus deactivating the equipment, activating the exhaust fan, and deactivating the makeup air being supplied for the common hazard area. Fire protection systems commonly have dry contacts that change state simultaneously with the discharge of the tanks. Those contacts trigger a gas valve and electrical shutoff response for the cooking equipment and provide notification to the building management system, alarm system, and other fire and life safety responses. Fire protection coverage over solid-fuel appliances requires additional consideration and needs to be applied based on manufacturer listing requirements and

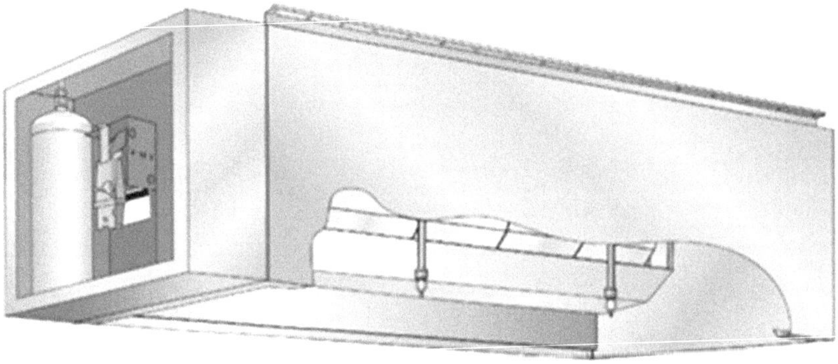

Figure 3.7 Fire protection tank.

Courtesy of Gaylord Industries

NFPA 96 (2021b). Note these systems must be isolated from other Type I exhaust systems to generate their own common hazard.

Grease filters. Grease removal devices are a foundational requirement of UL 710 (2012). Grease filters serve to remove larger particulate and protect the hood plenums and downstream ductwork from equipment fires moving up and through the exhaust system where they may spread to the rest of the building. Filters listed for use in generic hood systems are covered under UL 1046 (2010c), while permanent or removable grease removal devices integral to a listed exhaust hood are covered as a recognized component to the hood's listing. The performance of these filters is covered under ASTM F2519 (2020b) and VDI 2052 (2022) and is presented as "fractional efficiency." See Figure 3.8 for an example plot. The VDI standard also includes a loading-based method of test (MOT) used to determine how much grease is captured and what is reentrained back into the airflow. Note that filtration performance is not addressed during the UL 1046 (2010c) or UL 710 (2012) listing processes. Grease filters mostly use inertial impaction as the means to remove grease, with higher efficiency generally linked to higher static pressure and fan motor horsepower (kilowatts). Figure 3.9 shows different filter styles that have different efficiencies of grease capture at the hood. Designers should weigh the cost in motor horsepower (kilowatts) against a higher rate of grease deposition inside the downstream Type I ducts. Factors such as equipment type, volume, product cooked, duct length, local labor costs, and accessibility all impact the expense associated with maintaining Type I ductwork and, ultimately, the total cost of ownership. Greater detail on particle size and distribution generated from select cooking equipment and product types can be found in ASHRAE RP-745 (Gerstler et al. 1998) and ASHRAE RP-1375 (Kuehn et al. 2008).

Clean-in-place (water wash) and continuously operating mist hoods. Some manufacturers provide optional clean-in-place or continuously operating mist options for Type I hoods. Clean-in-place hoods typically introduce hot water and detergent through nozzles located inside the hood to clean away collected grease from the day's cooking operations. Such equipment can reduce or eliminate the need for dish-machine servicing of grease filters and the associated labor for the removal and replacement of the grease filters in the hoods. These savings come at the cost of the additional hot water, detergent, and sewer infrastructure needed to support the clean-in-place operation. Upsized water heaters and wash control cabinets are common additional requirements. Clean-in-place water consumption varies from about 0.5 to 1.5 gpm per linear foot of hood (6.2 to 19.3 L/s per linear meter of hood) based on hood type and manufacturer, with wash lengths lasting between 3 and 10 minutes per hood section. Continuously operating mist hoods are another option that are sometimes incorporated over solid fuel to mitigate the risks associated with burning wood or charcoal and the associated production of creosote and large quantities of sparks and embers.

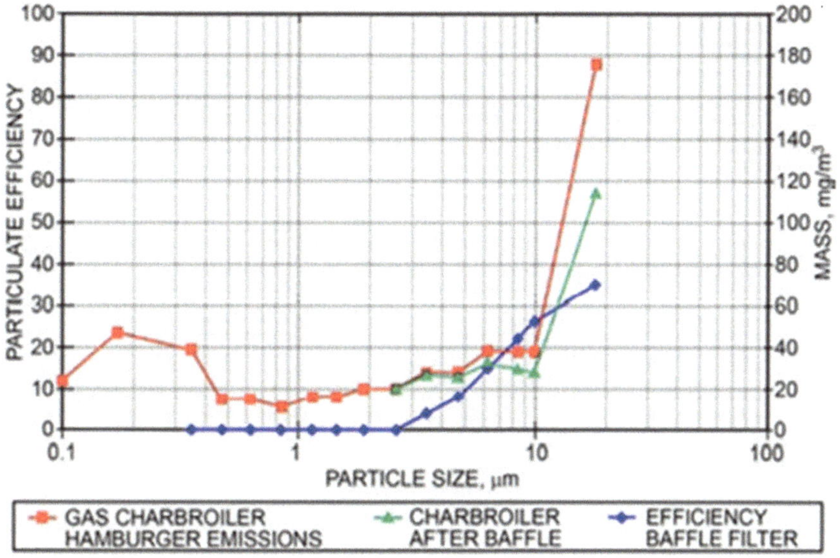

Figure 3.8 Grease filter fractional efficiency plot.
Courtesy of ASHRAE (2019)

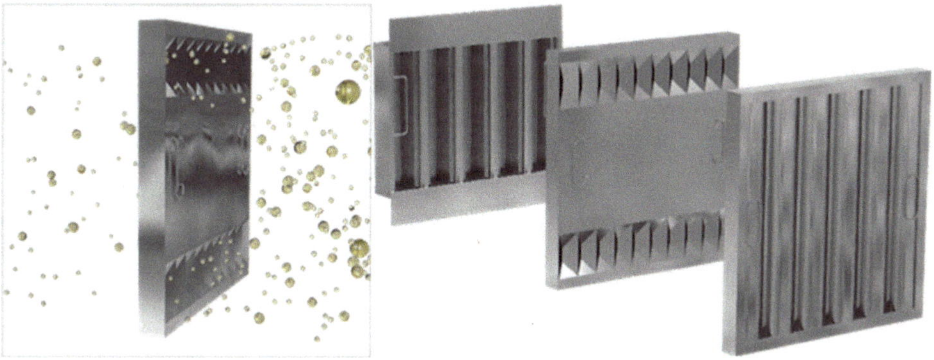

Figure 3.9 Examples of grease filter designs at increasing grease efficiency as well as increasing static pressure.
Courtesy of Accurex, a Greenheck Group Company

Additionally, misting can be used to bring condensable vapors out of the air where they can be mechanically filtered.

Canopy lights. Canopy hoods are overall the most common hood style in commercial kitchens, as they have listed lights installed inside the hood's reservoir to provide adequate light for safe food preparation. These light fixtures (luminaires) must be specifically rated for use in a commercial kitchen hood and are covered under UL 1598 (2021a). Available canopy lighting options range from incandescents to light-emitting diodes (LEDs). Initial cost and energy consumption must be balanced by the designer and applied appropriately to the hoods. Caution must be applied to large kitchens with incandescent or inefficient canopy lighting systems, as they may place a sizable electrical load on the kitchen. Note 50 fc (538 lm) at the cooking surface is a standard requirement for

U.S.-based kitchens and must be used as the basis for the design process to meet sanitation requirements in most localities.

DCKV controls. The *International Energy Conservation Code* (IECC; ICC 2021c) requires energy-reducing equipment such as:

- Demand-controlled kitchen ventilation (DCKV) on 75% or more of the hoods
- Heat recovery
- Maximizing transfer air that would normally be exhausted, to be provided on kitchens with total hood (Types I and II) exhaust rates of 5000 cfm (2360 L/s) or more

DCKV is a common energy-saving control used to meet code and reduce energy consumption, providing substantial fan law type energy savings on both the supply and replacement air sides as well as a reduction in the required tempering/conditioning needed to maintain space comfort. These controls primarily use temperature in combination with infrared (IR), opacity, or other sensing methods to determine the activity level of the cooking process and adjust the kitchen's exhaust fans to maintain C&C. DCKV controls are required to modulate both supply and exhaust airflows to the kitchen to maintain proper space pressurization and are sometimes provided with exhaust duct collar installed modulating dampers and pressure transducers to provide independent control of each hood section where multiple hoods are being served by a single exhaust fan. Controls of this type are typically listed in UL 710 (2012), UL 508A (2022b), UL 8782 (2017c), and more. See chapter 34 of *ASHRAE Handbook—HVAC Applications* (2019) for more information on applying DCKV systems.

Fire dampers. UL 710 (2012) allows for the listing of Type I hoods with or without fire dampers. Though once a standard in commercial kitchen hoods, they are seldom provided today due to very conservative and well-regulated design requirements for Type I ducts and effective fire suppression systems. Hoods with integrated fire dampers are listed for such use under UL 710 and are typically actuated by thermostats or fuse links.

Perimeter plenums and compensating hoods. Typical methods of managing the high volume of replacement (makeup) airflow required in commercial kitchens is through the use of perimeter supply plenums and, sometimes, compensating hoods. These components and features provide the means to introduce the airflow local to the hood while reducing the detrimental effects the makeup air can have on hood C&C. It is recommended that 50% of the makeup air delivered through perimeter plenums, shown in Figure 3.10, moves directly into the exhaust hood, while 50% of the load is introduced to the space and requires consideration in the kitchen's load calculations.

Many of the negative influences can be managed by reducing the volume and velocity of the airflow being introduced. Perimeter plenums, shown in Figure 3.10, when producing no greater than 75 fpm (0.381 m/s) on the downward edge of the hoods, have demonstrated the best overall practically applied performance. Refer to chapter 34 of *ASHRAE Handbook—HVAC Applications* (2019) for additional information. Perimeter plenums deliver makeup airflow continuously across each surface of the hoods they are installed against. Discharge temperatures exceeding 7°F (4°C) above ambient have been found to adversely affect hood performance and must be considered in the design process.

Compensating hoods, shown in Figure 3.11, allow for the introduction of the makeup air directly into the hood, minimizing the required space for makeup air inlets inside the kitchen. Contained inside the hood's product listing under UL 710 (2012), each of the systems shown have performance limitations and ratings needing to be closely followed to both meet code and perform. Note that internal discharge methodology is not accepted by ASHRAE Standard 154 (2022d) or the model mechanical codes at volumes exceeding 10%. Hoods properly producing internal discharge rates of 10% or less have been known to produce a reduction in the required exhaust rate.

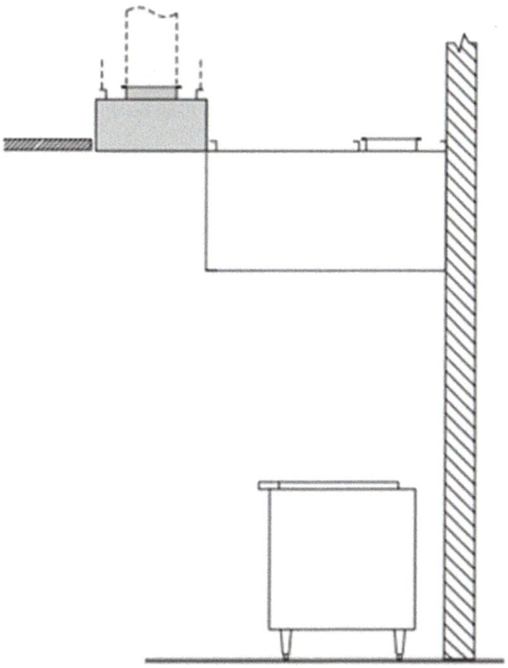

Figure 3.10 Perimeter supply plenum.
Courtesy of Gaylord Industries

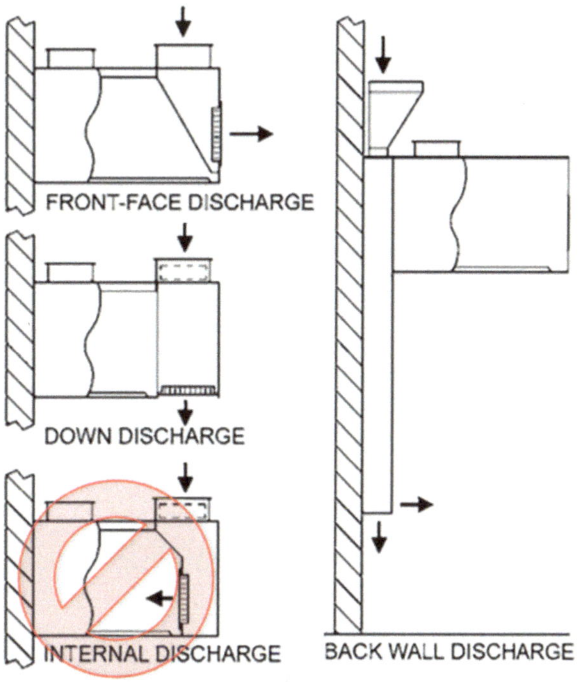

Figure 3.11 Common compensating hood designs.
Courtesy of ASHRAE (2019)

UVC lamps. Type I hoods can be provided with integrated UVC systems used for the reduction of deposited grease and possible reduction in cooking odor. Hoods with listed UVC systems typically install lamps in the path of the exhaust allowing for direct exposure of the airborne grease to UV light and, additionally, the downstream subjugation of the grease duct to ozone produced by the lamps. Such systems require safety interlocks to eliminate potential exposure of persons to UVC light and ozone and are certified under UL 710C (2006). Field and laboratory efficacy studies funded by manufacturers, ASHRAE, and other organizations have been conducted with varied results in quantifying the effectiveness of UVC to reduce deposition both in real time and as a treatment. There is currently no publicly accepted performance standard associated with UVC and its grease-reducing capabilities in Type I ductwork.

Other subassemblies. Ozone generators, electrostatic cells, integrated UL 900 filters, odor mitigating chemical spray, water misting, and other systems are available from manufacturers to address smoke, odor, and Type I system maintenance. Verify with providers the equipment is listed for such use and, if available, consider all sources of performance data in your selection to ensure the desired intent is being met. See chapter 4, "Design Considerations," for additional information.

Type I Hoods—Ratings

UL 710 (2012) is the basis for Type I hood ratings. Hood electrical ratings, front and side overhangs, mounting heights relative to the cooking surface, non-continuously welded joints, clearance to combustibles, equipment duty coverages, exhaust collar position ranges, presence or absence of fire damper, grease removal devices, and minimum ventilation rates all come from the product's listing. Type I hoods come marked with the electrical rating, presence or absence of damper, equipment duty (cooking temperature), section airflow, and static pressure from the manufacturer on exhaust-only hoods. Compensating hoods also have design airflows for the hood's makeup air inlet. These values are typically unique to each hood section.

Reduced clearance to combustibles. Some Type I hoods have reduced clearance to combustible listings, reducing the installed distance from the standard distances specified inside the model mechanical codes. These ratings are achieved by a provision inside UL 710 coming from UL 710B (2011a) and UL 197 (2010b), which allows the reduction in space between combustibles and heated surfaces when the equipment is confirmed experimentally to not exceed 117°F (48°C) during normal operations or 175°F (80°C) during abnormal conditions.

Duty level (solid fuel). Another rating for Type I hoods is the duty level of the equipment being covered. Typical Type I hoods cover light to heavy duty, with some being designed for use over solid fuel. Previously, the solid fuel rating under UL 710 was designated by a cooking surface temperature of 700°F (371°C). The current UL duty rating for solid fuel is "extra heavy," which is the designation for a hood installed over equipment obtaining all or part of its input energy from the burning of charcoal or wood. The uncontrolled nature of this type of cooking equipment requires special attention regarding the design and ratings of the Type I exhaust system. Solid-fuel-rated grease filters, higher ventilation rates, isolated fire suppression systems, and separate exhaust systems must be used to comply with the model mechanical codes. Equipment rated as "extra heavy duty" requires isolation from electric- and natural gas-heated appliances, except where the lesser-duty equipment sharing the hood does not require surface fire suppression. When hoods supporting solid fuels are used in line with other lesser-duty hoods, the installation of full end panels is required where hood sections meet. Some of these hoods may be provided with misting nozzles to reduce the risk of fire in the downstream ductwork. Refer to specific filter, hood, and fire protection manufacturer listings for compliant applications and solutions.

Type I Hoods—Hood Styles

Commercial kitchen hoods come in several styles, all of which impact integration of the hoods into the building, the amount of space they require, and their demand for fresh air. The specification of one style over another is determined by the cooking equipment's type and location inside the kitchen. The most common hood styles in use today are wall-mounted canopy hoods because of their versatility in operating over tabletop equipment as well as larger equipment such as ovens, combi ovens, rotisseries, and more. Figure 3.12 shows common Type I canopy hood styles.

Canopy hoods can be mounted against code-compliant limited or noncombustible walls or in island applications where all sides are open. They are sometimes provided with incorporated makeup air plenums; these are referred to as *compensating hoods*. Compensating hoods introduce makeup air horizontally away from the hood, downward—typically right over the chef's head—or internally into the hood's capture reservoir, where volumes greater than 10% are not accepted (see Figure 3.11). Typical canopy hoods are designed to hang from the ceiling, and their weight must be accounted for in the structural design, following mechanical code-required clearance to combustibles. Canopy hoods are the most commonly used hoods over solid-fuel appliances, where they are supported by independent fire protection and exhaust systems. Disadvantages of canopy hoods are

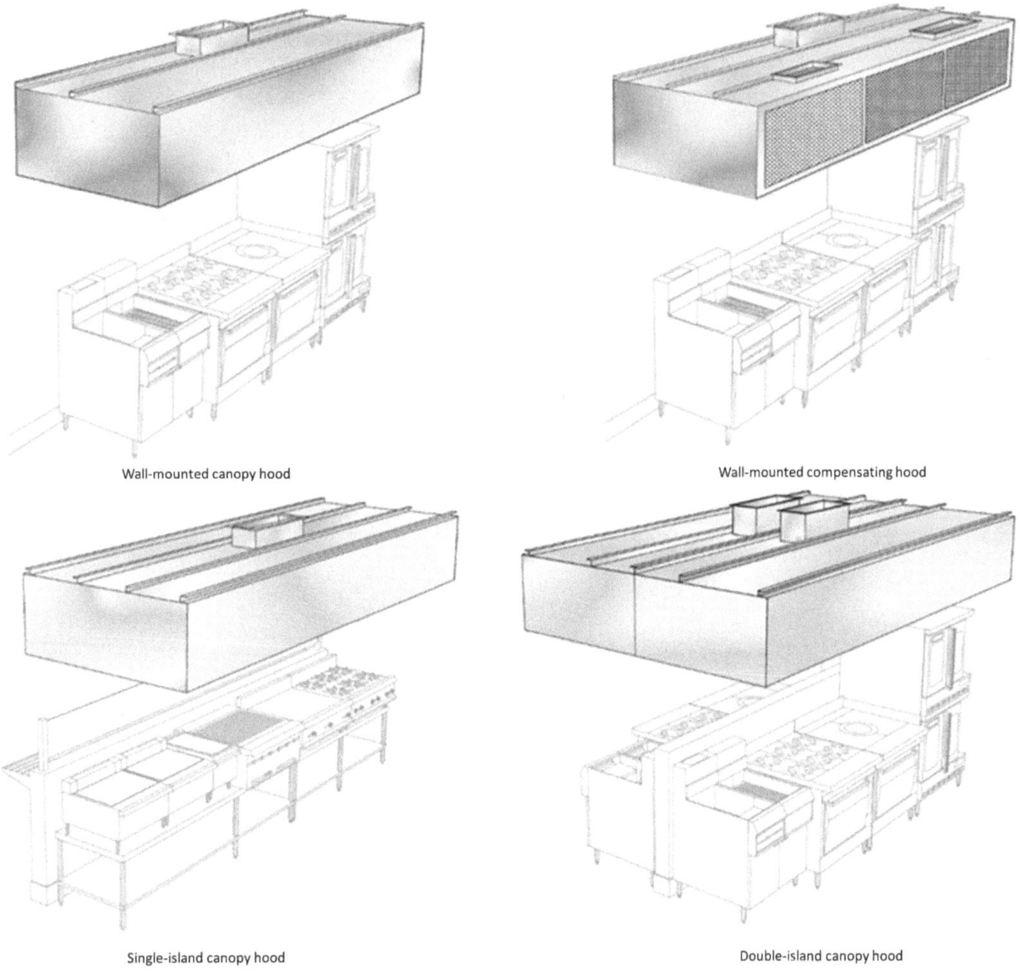

Figure 3.12 Type I canopy hood styles.

their slightly higher airflow requirements compared to more specialized hoods such as backshelf, eyebrow, or passover style hoods.

Backshelf style hoods, shown in Figure 3.13, are commonly used in quick-service or other specialized applications where equipment is predominately countertop. The reduced distance from the cooking surface to the inlet of the filters reduces the size of the cooking equipment's plume, thereby reducing the required exhaust ventilation rates needed to achieve C&C. ASHRAE RP-1202 (Swierczyna et al., 2005) has shown has shown that for each inch of hood mounting height reduction, a resulting 1% reduction in the exhaust rate was achieved.

Eyebrow hoods, shown in Figure 3.14, are commonly used in applications where large deck, baking, or pizza ovens are used. They are designed around direct mounting to the oven or equipment they serve and are sometimes available with direct venting for the equipment's flue as well as small canopies for capturing emissions from operable doors or openings. Eyebrow hoods can reduce ventilation rates and the space required in the facility's kitchen compared to that of canopy style hoods.

Less common styles such as downdraft, ventilated ceiling, proximity, and passover hoods are also available, each providing optimized conditions for specific equipment types or locations inside the facility's kitchen or dining spaces.

Other available solutions are ventless, or recirculating, hoods. These hoods allow for the treatment of grease-laden airflows requiring a Type I system without the need for the exhaust duct, shafts, fans, and related systems. These systems are for use over electric equipment and do not commonly address the sensible or latent heat generated by the cooking process, so these heat gains must be accounted for in the HVAC system design. They typically rely on a series of mechanical, electrostatic, and gas-phase filters to reduce emissions to acceptable levels at the point of discharge back into the space. Careful consideration is recommended for the impacts of ventless hoods on the facility's indoor environmental quality (IEQ). These hoods are listed to UL 710B (2011a) and are provided with the means to reduce the emissions to below 3.1×10^{-7} lb/ft^3 at 500 cfm (5.0 mg/m^3 at 236 L/s) (ICC 2021d). EPA Method 202 (2016) is used as the basis for emissions testing done under the UL 710B (2011a) evaluation. While the code does not explicitly prohibit it, installing multiple recirculating hoods whose combined operations would exceed the 3.1×10^{-7} lb/ft^3 at 500 cfm (5.0 mg/m^3 at 236 L/s) limit is not recommended.

Properly analyzing, reviewing and selecting a Type I hood style can ultimately affect the required ventilation rates needed for the kitchen. It is the authors' opinion that significant airflow savings will result from specifying backshelf hoods in place of canopy hoods or wall-mounted canopy hoods in place of single-island hoods. These airflow reductions result in not only savings to the size of the exhaust fans but also the Type I grease duct(s), shaft(s), and comfort HVAC replacement air systems. Savings can account for thousands of dollars in capital and annual operating costs. See chapter 1 for assessing life-cycle costs and total cost of ownership.

Type I Hoods—Determining Hood Airflow

Type I hoods are designed to capture, contain, and manage the grease and threat of fire generated by the equipment over which they are placed. Fulfilling this objective efficiently while meeting the energy and mechanical codes and the product's listing can be a challenge. For the purposes of this guide, only "listed" Type I hoods are covered due to their specification being the only way to meet both the model mechanical and energy codes simultaneously.

Meeting code. The IECC (ICC 2021c), based on ANSI/ASHRAE/IES Standard 90.1 (ASHRAE 2022b) for commercial buildings, sets maximum ventilation rates for generic Type I hoods based on the highest-duty equipment rating located beneath each hood section. See Figure 3.1 to start the process of determining hood type and duty. As of 2020, nearly 40% of U.S. states utilize the maximum airflow cap requirements designated in IECC, with other states soon to follow. With this in mind, it is

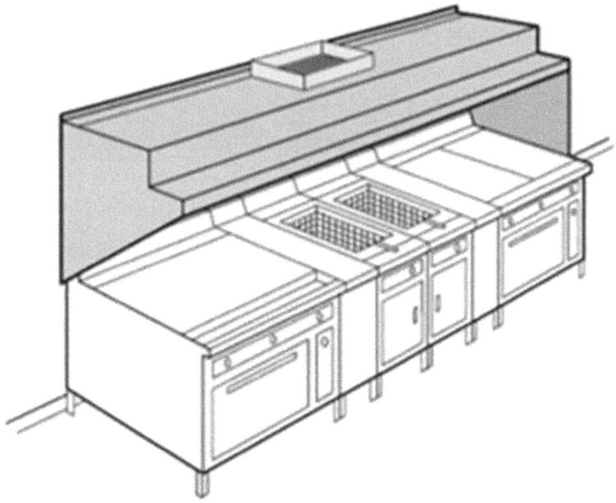

Figure 3.13 Backshelf hood.

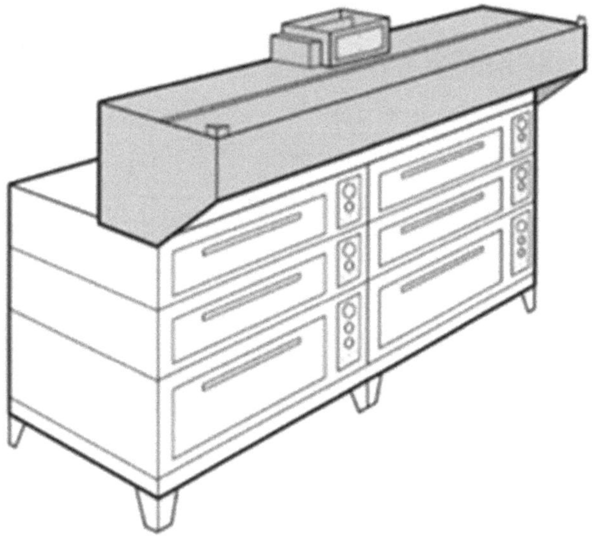

Figure 3.14 Eyebrow hood.

recommended that hood ventilation rates not exceed ASHRAE/IES Standard 90.1 ventilation rates to avoid code exceptions to achieve permitting.

Minimum exhaust flow rates are set by each hood's listing. UL 710 (2012) contains a method of test called "Cooking Smoke and Flare up," which is used to determine minimum ventilation rates for each hood section operating at the longest and shortest available section lengths. The testing is done generically with a hamburger on fully heated appliances representational of light, medium, heavy, and extra-heavy-duty cooking lines. In older versions of UL 710, the duty levels were represented by cooking surface temperatures. A representative of a nationally recognized testing laboratory (NRTL) is commonly used to administer the testing in a draft-free laboratory environment. The resulting values are published and show minimum ventilation rates alongside their corresponding equipment duties, overhangs, and hood heights relative to the cooking surface.

Hoods designed at or above their product's listed values are accepted as meeting code pending a field evaluation; however, the minimum listed ventilation rates published are often inadequate to fully capture and contain all equipment emissions and therefore must be used only as a floor for design exhaust rates and not the basis of design. Only under circumstances in which equipment placement, diversity, and features are incorporated into the design can exhaust rates drop to within listed rates. Be sure to consult hood manufacturers for proper design exhaust rates and opportunities for reductions in those rates. See chapter 8, "TAB and System Commissioning," for details on the commissioning and field evaluation of a hood's performance.

Design considerations/best-practice design. Once the operational ranges to meet code have been determined, setting airflows relies on several factors, including hood selection and type, equipment position and diversity, overhangs (hood sizing), replacement air sources, and sometimes the addition of accessories such as side panels and rear seals. Many of these factors' effects on hood performance have been documented through ASHRAE research such as RP-1202 (Swierczyna et al. 2005) and RP-1480 (Swierczyna et al. 2010). Both studies used ASTM F1704 (2022) to determine the effects of equipment position, diversity, side panels, rear seals, and replacement air introduction on hood performance. ASTM F1704 makes use of enhanced visualization methods such as Schlieren and shadowgraphs to determine minimum hood capture airflow rates under controlled conditions.

Results from ASHRAE RP-1202 (Swierczyna et al. 2005) indicate that on a generic hood, when best practice in all categories is observed, up to 45% of the required exhaust can be saved—a net 2300 cfm (1086 L/s) in savings off a 10 ft (3.05 m) generic hood. These best practices include the following:

- Moving heavy-duty equipment toward the center of the hood on mixed-duty lines, increasing hood front and side overhangs
- Minimizing mounting heights to not exceed those prescribed by the Americans with Disabilities Act (ADA 2023)
- Pushing heavy equipment against the rear wall and adding rear seals between the wall and the heavy equipment
- Adding side panels

In the experiments, all of these actions produced substantial reductions in the required exhaust ventilation rates.

ASHRAE RP-1480 (Swierczyna et al. 2010) further studied best-practice design on island hoods, which shared many of the same benefits for ventilation rate reductions with increased front and side overhangs and reduced turbulence from low-velocity-introduced replacement air as wall-mounted hoods. RP-1480 showed that single-island applications benefit from a V-back design compared to single-sided filter configurations, confirmed that double-island equipment lines performed similarly to wall hoods regarding ventilation rates and that introduction of air in excess of 10°F (5.6°C) to the kitchen's ambient temperature does impact hood performance, and further illustrated how critical a balanced introduction of replacement air is to island hoods. To put it simply, replacement air is to be evenly distributed around the hood, at low velocity, and as far as possible from the hood. Perforated perimeter plenums installed on all sides, introducing replacement air at 80 fpm (0.406 m/s), outperformed other strategies in the research. When considering all options, perforated plenums are the recommended methodology to introduce makeup air without jeopardizing hood performance.

Setting Type I hood ventilation rates. The best source for setting ventilation rates for Type I hoods is the equipment/hood manufacturer. They are versed in the equipment's UL and ASTM F1704 (2022) minimum airflow results and understand the impacts of the introduction of makeup air, equipment diversity, equipment position, hood style, type, and other features on determining comfortable and efficient ventilation rates. Many manufacturers have automated processes or software to assist in generating hood selection, sizing, and design airflows. These methods are typically empirically based and sometimes use plume-size calculations coupled with environmental factors to determine airflows. Utilizing manufacturer methods for design is recommended and has benefits not only for determining optimal airflows but also in delivering accountability. Similar to other industries, there are performance, maintenance, and upfront cost benefits to be had with each manufacturer's Type I hoods, each of which need to be weighed in the design process.

In addition to manufacturer-published UL and ASTM F1704 (2022) values, confirmation of manufacturer design airflows is available in the appendix of ASHRAE Standard 154 (2022d). It uses a method sourced from the German standard VDI 2052 (2022) and determines hood airflow by summing each piece of the equipment's plume, accounting for the introduction of replacement air, hood style, performance, and type. Refer to ASHRAE Standard 154, Informative Annex B (2022d), for a simplistic approach and unlisted hood example. It is recommended that the method only be used as a reference, because each manufacturer's hood performance impacts constants used in the calculation.

Effects of replacement air on hood performance. A major factor that cannot be overlooked in the successful design of a high-performance CKV system is the effects of replacement (makeup) air on hood performance, IEQ, and space comfort. A successful replacement air system must be designed to deliver air that will not adversely affect the space comfort or hood performance. Replacement air must be delivered equally on all sides and at low velocity. The introduction of replacement air must not induce velocities greater than 75 fpm (0.381 m/s) at the lowermost portion of the kitchen's hoods. All sources of transfer air, such as pass-through windows and/or doors, should not exceed 50 fpm (0.254 m/s). Typical hood design relies in part on the fact that the turbulence inside the kitchen will be at a minimum, with single-island applications being the most sensitive to a turbulent condition.

Commissioning. Verification is a critical part of successfully providing a high-performance kitchen. Once hoods are installed, their performance must be verified operationally to the method of test spelled out in ASHRAE Standard 154 (2022d). The hoods and all sources of replacement air are to be balanced to design conditions, and a visual confirmation of each hood's successful C&C is to be completed. This process must be accomplished over fully heated cooking surfaces under operational conditions. All sources of replacement air are to be active throughout the process and operating inside of 0–0.02 in. of water (5.0 Pa). Refer to chapter 8, "TAB and System Commissioning," for details.

Type I Hoods—Design Approach

The role of the food service consultant is a critical first step in the design process. They determine the facility layout and equipment ratings based on the menu, volume, type, and occupancy desired by the owner. The equipment position, type, size, and layout are drawn and specified. This information is then used as the basis to select and size the hood(s). The resulting hood engineering is used in the base calculations for the mechanical support system design. The process of specifying the hoods comes from the project's food service consultant or mechanical engineer, both of whom will use manufacturer's engineering or automation process to support the design of the project's listed hoods, fire protection, and downstream Type I exhaust systems. It is not uncommon for hood design to come from the consultant, with the remaining system components designed by the mechanical engineer. Design crossover between the food service consultant and the mechanical engineer will happen. Coordination between the two designs is critical to ensure all requirements are met, with the engineer of record (EOR) needing to provide final design approval.

Replacement air systems are sized using local climate zone data, exhaust airflows, and comfort set points. The project's mechanical engineer then designs the replacement air system to meet code and maintain facility comfort, pressurization, and energy use optimization. Using all sources of facility transfer air and the balanced, low-velocity introduction of that air into the kitchen space is critical to delivering a high-performance kitchen hood system. See chapter 8 for more details.

Type I Hoods—Ventilation System Supplemental Equipment

Additional support systems exist for Type I hood systems, including the following:

- PCUs used to abate cooking emissions and odor
- Heat recovery systems used to reuse the heat generated from the cooking process for the heating of replacement air or facility hot water
- Utility distribution systems used to safely route electrical power and resources such as hot water and steam to commercial cooking equipment

These listed systems allow for purpose-built and sometimes capitalizable solutions to meet energy code requirements and emissions standards, optimize the total cost of ownership, or increase kitchen longevity. Consult *ASHRAE Handbook—HVAC Applications* (2019), chapter 34, "Kitchen Ventilation," and chapter 4 of this guide for more information on equipment deliverables.

4

Design Considerations

Commercial kitchens are a complex system of integrated components including, but not limited to, the cooking appliances and processes, exhaust ventilation hoods, ductwork, exhaust fans, replacement air systems, supply air diffusers, heat recovery systems, safety systems, refrigeration systems, control systems that regulate how these various components interact with each other, and energy management systems. The overarching purpose of the commercial kitchen ventilation (CKV) system is to remove heat, smoke, steam, grease emissions, and combustion byproducts from the kitchen in order to provide a safe environment with acceptable indoor environmental quality (IEQ) for the cooking staff. Additionally, pollution control units (PCUs) can be installed downstream of the hoods to manage the emissions being discharged to the atmosphere. Figure 4.1 gives an overview of the design process discussed in this chapter.

Types of Cooking Processes and Appliances

The starting point for every commercial kitchen design is identifying the cooking processes that will be used and the cooking appliances needed to perform those operations. Some of the primary cooking processes in commercial kitchens are presented in Table 4.1.

Once the cooking operation has been determined, the cooking appliances need to be selected. Note that there may be several different types of appliances that can perform a given cooking operation. For example, if a cook wanted to roast meat, the choice of appliances could include a standard oven, convection oven, or combi oven. The differences between these appliances are as follows:

- A standard oven encompasses the food with hot air
- A convection oven circulates the hot air, which speeds up the cooking
- A combi oven allows the addition of steam, which can help prevent the food from drying out during the cooking process

Additionally, some cooking appliances can perform multiple cooking processes. For example, a combi oven can bake, roast, steam, proof, and rethermalize food. It is important to understand that, typically, the food service consultant performs the function of selecting the cooking appliances to meet the desired menu goals. The consultant selects and optimizes the quantity and physical size of the cooking equipment as well as energy consumption, both of which can lead to savings on the overall kitchen ventilation and HVAC systems. While the CKV designer may not be directly involved in the appliance selection, it is important that they understand how and why this process is one of the first steps taken in the design stages of a commercial kitchen.

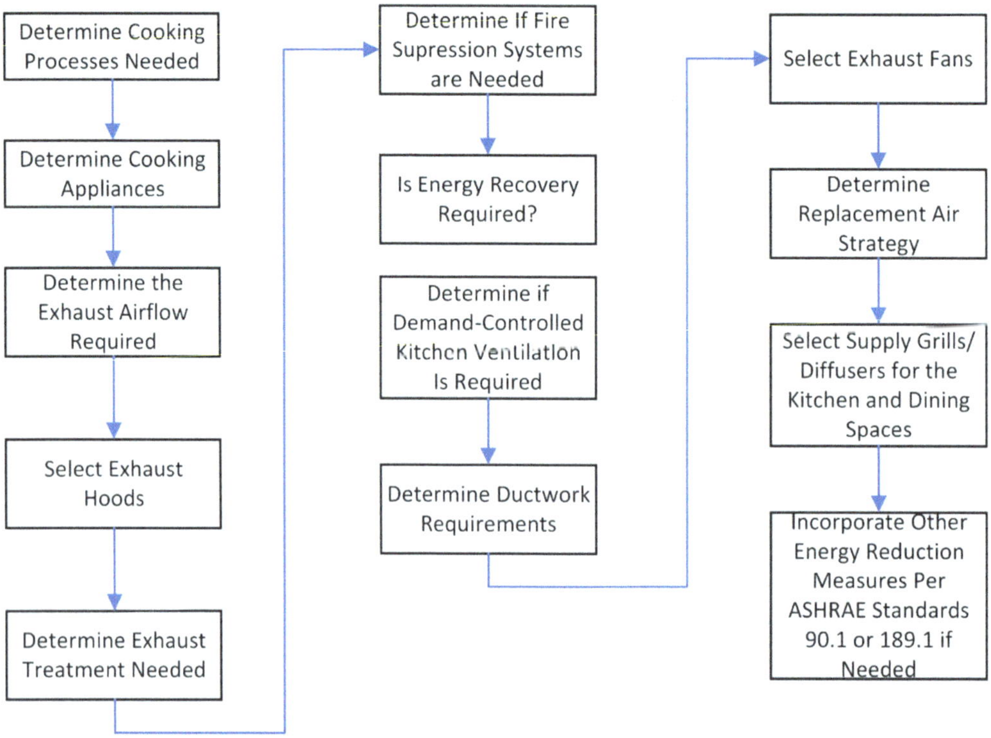

Figure 4.1 Commercial kitchen design process.

Table 4.1 Cooking Operations and Commonly Associated Appliances

Cooking Operation	Description	Typical Appliances
Roasting/baking	Food is cooked with hot air, which can be dry or have moisture added, that surrounds and heats the food.	Convection oven Combi oven Range oven Pizza oven
Broiling	Food is cooked by direct convective heat and possibly radiation energy.	Underfired broiler Overfired broiler Salamander broiler/grill
Surface or pan frying	Food is cooked on a hot surface that is coated with oil or fat.	Griddle Range top Hot top Wok range Induction range Plancha
Boiling	Food is cooked in boiling water.	Pasta cooker Range top
Steaming	Food is surrounded by steam to cook quickly.	Steamer Combi oven Steamer basket
Deep-fat frying	Food is cooked in a vat of oil.	Deep-fat fryer

Effluent Production by Cooking Process

It is important for the designer of the kitchen to understand that not all cooking process are the same—each one is unique with regard to what emissions and other byproducts are produced, as described in this section. When food is cooked, the following emissions are created to varying degrees, dependent upon the cooking process: grease, which consists of solid particles as well as gaseous vapor; smoke; steam; and other gases typically in the form of volatile organic compounds (VOCs). Note that the VOCs also carry much of the perceived odor that is created during the cooking process. Two ASHRAE-funded research projects, ASHRAE RP-745 (Gerstler et al. 1998) and RP-1375 (Kuehn et al. 2008), focus on grease emissions from larger hooded appliances. For these appliances, both the particulate grease emissions (by particle size range) and vapor emissions are presented in Figure 4.2. It is evident from Figure 4.2 that the cooking emissions vary largely as a function of the cooking process. The two processes with the largest overall grease emissions include using a solid fuel (wood-fired) broiler to cook hamburgers, which has total emissions of approximately 140 lb/1000 lb (140 kg/1000 kg) of food cooked, and using a Chinese wok to fry chicken, which has total emissions of nearly 250 lb/1000 lb (250 kg/1000 kg). Contrast this with the emissions from using a gas fryer to cook french fries, where the total measured emissions were less than 2 lb/1000 lb (2 kg/1000 kg), which is less than 1% of the emissions from the Chinese wok.

Figure 4.3 presents the breakdown of grease emissions by cooking operation into what percentage of the emissions is grease particulate versus grease vapor. This information becomes valuable when selecting exhaust filtration in the hood along with other treatment systems, because grease vapor will travel through mechanical grease filters and separators. It is particularly important for the engineer and end user to understand that mechanical filters (and media filters) generally only collect grease particulate unless the surfaces are cold enough to provide condensation of the grease vapor.

Another important component of cooking emissions is smoke, especially when cooking proteins on a hot surface (such as hamburgers being cooked on an underfired broiler) or when fuel (such as charcoal or wood) is being combusted during the cooking process. One group of researchers focused on the emissions from different types of wood being combusted. They found that for oak wood in particular, the largest mass of emissions was at a particle size of 0.1 microns, which is smaller than what they found for charbroiling meat, which had the largest mass at a particle size of 0.2 microns (Kleeman et al. 1999). From a designer's perspective, it is important to know what particle sizes smoke comprises and whether smoke mitigation is a requirement for the project.

Effluent from Countertop Appliances

ASHRAE RP-1631 (Zheng et al. 2016) evaluated the grease emissions from smaller countertop appliances. Figure 4.4 shows the total grease concentrations emitted from these appliances in lb/ft^3 (mg/m^3). It can be observed that the emissions from the hot plate, rapid-cook oven, and hot dog roller all exceed the 3.12 lb/ft^3 (5 mg/m^3) concentration limit where hoods may be required by the local authority having jurisdiction (AHJ). *International Mechanical Code* (IMC; ICC 2021d) states that if an electric cooking appliance produces 5 mg/m^3 (3.12 lb/ft^3) or less of grease when tested at an exhaust airflow rate of 500 cfm (850 m^3/h), it does not need a hood. Figure 4.5 presents the comparison of particulate and vapor emissions as a percentage of total emissions for these same countertop appliances. When evaluating a kitchen, the designer should consider grouping countertop appliances together underneath a Type I or Type II hood.

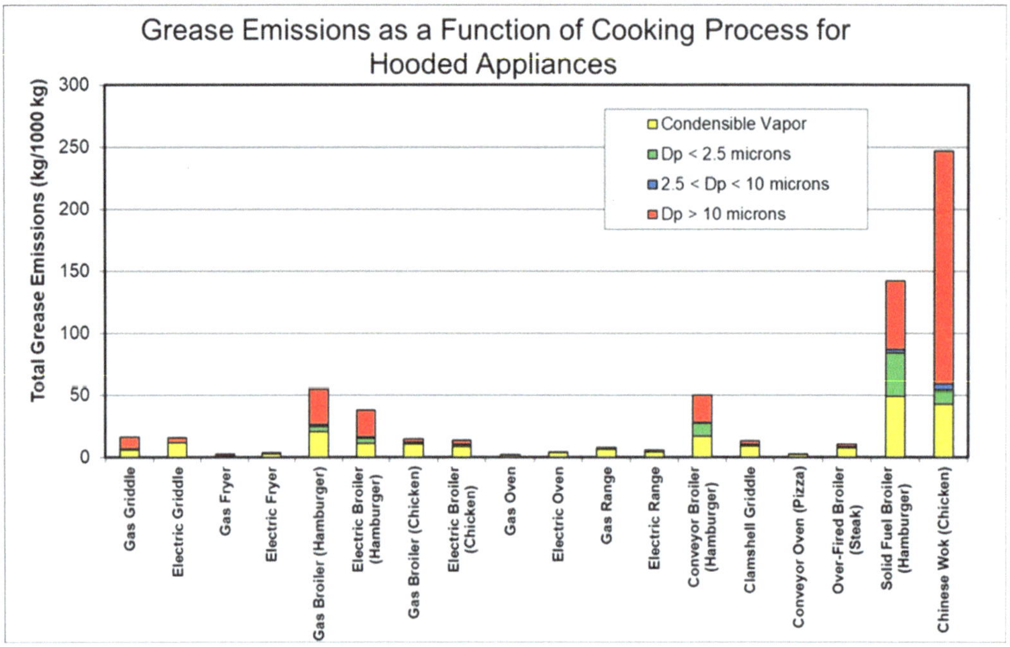

Figure 4.2 Grease emissions from hooded cooking processes.
Courtesy of ASHRAE (Kuehn et al. 2008)

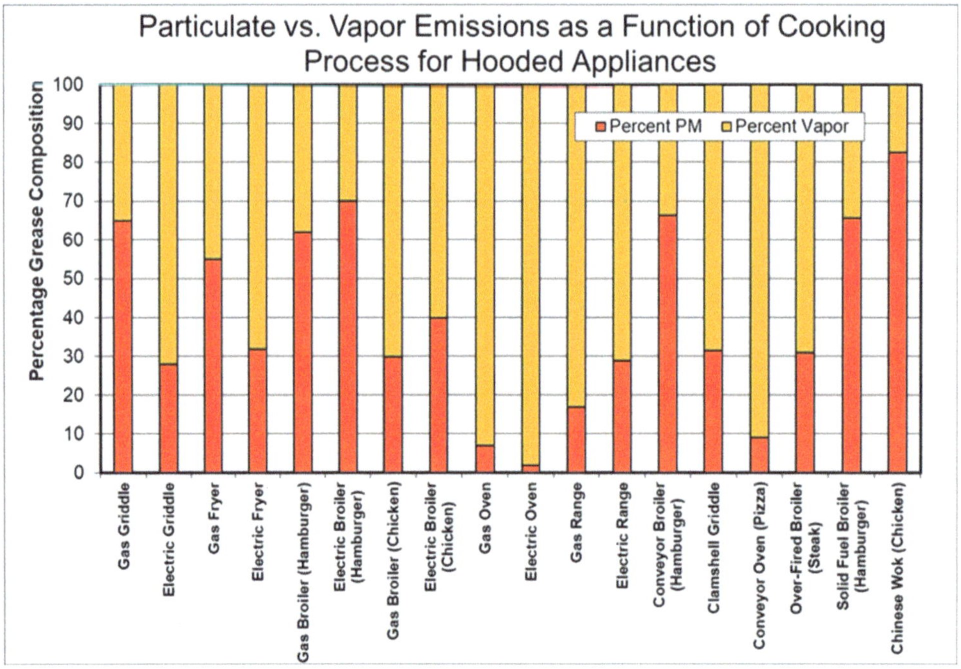

Figure 4.3 Grease particulate matter (PM) and vapor percentages from hooded cooking processes.

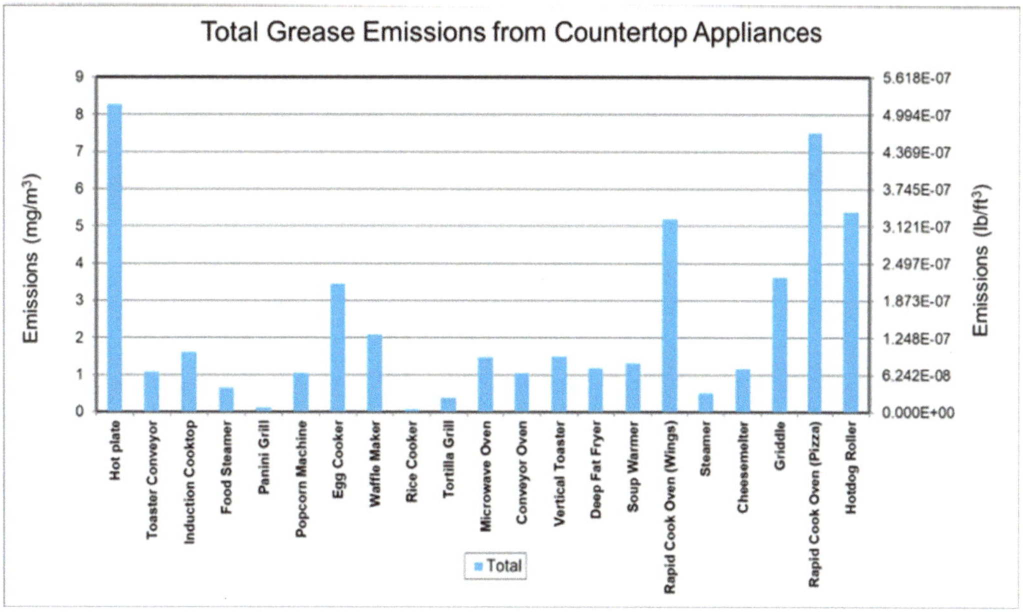

Figure 4.4 Total grease emissions from countertop appliances.

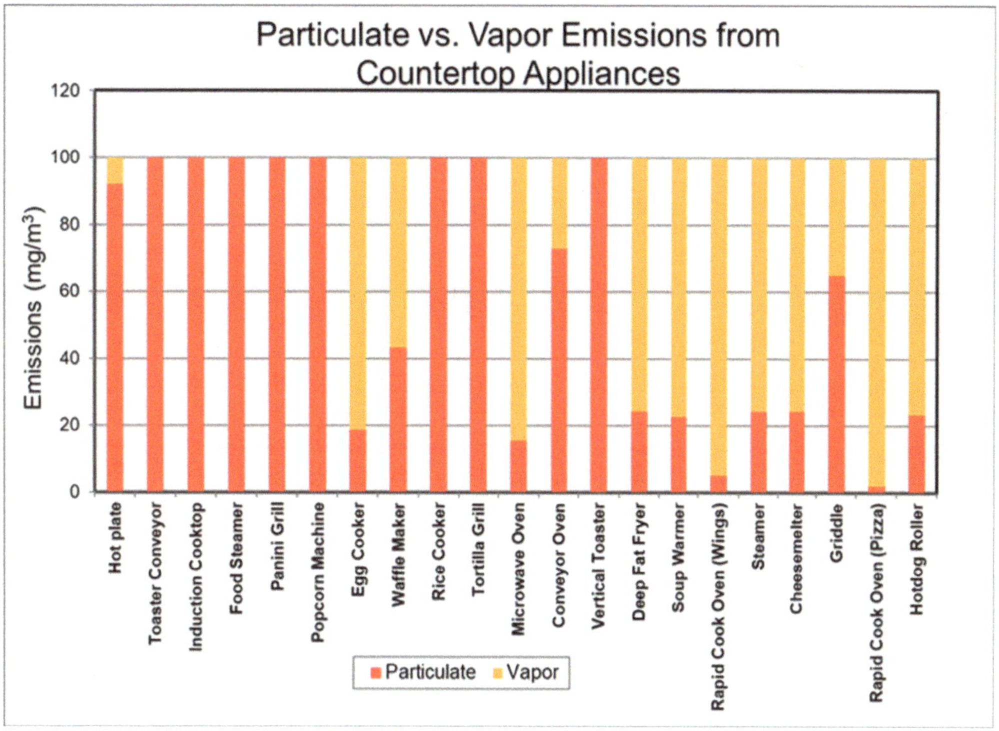

Figure 4.5 Grease particulate and vapor percentages from countertop appliances.

Types of Exhaust Hoods

Chapter 3, "Exhaust Hoods and Related Equipment," provides detailed descriptions of the types of hoods used in commercial restaurants, the duty levels of the cooking equipment, and the design processes for Type I and Type II hoods.

To summarize chapter 3, for the purposes of this chapter, there are two broad types of exhaust hoods: Type I and Type II. A Type I hood is required when the cooking processes produce grease effluent. A Type II hood is required when an appliance is producing only heat or moisture (no grease or smoke is being produced). The most common application for Type II hoods is installation over dishwashers or steam-producing equipment. Both of these hood types are connected to the building's exterior through the use of ductwork and exhaust fans.

Hood selection is crucial from energy and hood performance perspectives. Two means of improving a hood's energy performance are moving the extraction point closer to the cooking surface and minimizing entrainment air through means such as end panels. Hood performance can be impacted by making sure the correct hood is selected for a given application. Factors such as proper overhang, adequate grease filtration, and others should also be considered. Additionally, many hood manufacturers have design features that enhance their products' hood efficiency.

Because Type II hoods exhaust air that is predominantly free from smoke and grease, there is not a reason to further clean this air; however, the exhaust from Type I hoods can contain large quantities of grease and other pollutants that should be removed from the airstream.

Recirculating/Ventless Systems

Recirculating, or ventless, systems are a form of ventilation in which the exhaust byproducts (including heat, grease, and smoke) are emitted back into the kitchen or building space at allowable levels. Chapter 3 also contains information on this topic. The two types of ventless hoods are non-integral and integral recirculating systems. Non-integral systems may be used with generic electric cooking equipment up to specified size and voltage requirements. Integral recirculating systems are designed for a specific make and model of cooking equipment.

Recirculating hoods that filter the exhaust air and recirculate that air back to the space are becoming more common in both non-kitchen and kitchen spaces. Two critical design aspects that need to be addressed when using ventless hoods are:

- Ensuring that the HVAC system is adequately sized to handle the added sensible, latent, and radiant loads (e.g., heat gains) to the space, because all of the energy going into the appliances ends up going to the food or as load to the space.
- Ensuring there are sufficient air changes in the space to help keep the air cleaner by supplying fresh air. This reduces the concentration of grease particulate, vapor, and odors in the space.

Another operational aspect of ventless systems is that maintenance will be required to clean or replace the filtering components. The maintenance requirements vary depending on the types of filtration used and the food products being cooked. Example schedules of some filtering devices are listed in Table 4.2 (refer to manufacturers' instructions for schedules on their specific systems).

Exhaust Treatment Systems

When it comes to treating exhaust from cooking processes, it is important to know what is being emitted, including grease particulates, grease vapor, smoke, and odors. The "Effluent Production by Cooking Process" section of this chapter discusses grease particulate matter and vapor and presents limited data on VOCs, which can contribute to odor in the emissions.

Table 4.2 Example Inspection and Cleaning Periods for Filters in Ventless Systems

Filter Type	Typical Inspection/Cleaning Schedule
Primary grease extractor	Daily
Secondary grease extractor	Daily
First-stage media filter	1 to 2 months
Electrostatic precipitator	Daily to weekly
High-efficiency particulate air (HEPA) filter	2 to 3 months
Carbon filter	2 to 3 months

During the design process, which pollutants need to be managed and how to manage them need to be determined. The degree to which the effluent needs to be managed depends largely on two questions:

1. What local emission requirements need to be met?
2. Is the commercial kitchen located in an area that could make it susceptible to nuisance complaints from neighboring buildings or tenants?

A classic example of this situation is restaurants located on the bottom floor of condominiums, where smoke, odors, and grease can become nuisances for the tenants and building management.

Hood-Based Exhaust Treatment

There are several types of treatment systems commonly used in exhaust hoods that serve as the first line of defense in treating the effluent generated by cooking processes. All the technologies presented in Table 4.3 can be combined to address a specific situation, and while the suppliers should be consulted, the engineer needs to understand the application and verify that the supplier's recommendation will achieve the desired goals.

The most common types of systems installed in hoods include mechanical filtration, ultraviolet-C (UVC) lamps, water mist systems, water wash systems, and electrostatic precipitators (ESPs). It is important to understand that these types of systems in hoods do not typically remove smoke. High-efficiency particulate air (HEPA) filters are the optimal control technology choice when nearly complete smoke removal is required. The primary reason for this is that, as discussed previously, smoke is typically in a size range of 0.1 to 0.2 microns, which is a region where mechanical and media filters have poor particulate efficiency, while a HEPA filter has a particulate removal efficiency of at least 99.97% in this particle size range. Refer to the "Media Filters" subsection of the "Pollution Control Exhaust Treatment" section of this chapter for more information on test standards and classifications. Table 4.3 presents a typical application matrix of when to use which technology, but the actual selections can vary depending on the customer's needs and cost considerations.

Mechanical filters. Mechanical filtration is included with all exhaust hoods for CKV applications that produce grease. NFPA 96 (2021b) requires that mechanical filters be listed as grease removal devices. These may either be removable or nonremovable (integrated) filter types. It is important to consider that the efficiency of the filter varies depending on the cooking process, the filter used, and the airflow at which the hood is operating. ASTM F2519 (2020b) evaluates the performance of mechanical filters specifically for use with CKV systems and shows the efficiency of the filter over a broad range of particulate sizes (from 1–10 or more microns). It is important to understand that mechanical filters do not remove grease vapor, only the particulate. An article in

Table 4.3 Typical Application Matrix Based on Cooking Duty

Cooking Duty Level	Mechanical Filtration	UVC Lamps	Water Mist	Water Wash	ESPs	HEPA	Carbon Filters	Ozone Generators
Light	x	x						
Medium	x	x		x	x			
Heavy	x	x		x	x			
Solid fuel	x		x	x	x		x	
Smoke control					x[1]	x		
Odor control		x					x	x[2]

[1]The degree of smoke removal varies with parameters such as the velocity of the air moving through the ESP cells and the number of passes that are installed.
[2]Ozone generators are less effective at deodorizing when proteins are being seared.

Table 4.4 Comparison of Particulate and Total Grease Removal for Three Grease Filters Used over a Gas Griddle Cooking Burgers

Filter	Total Filter Efficiency	Filter Efficiency for Particulate Only
A	60%	92%
B	22%	33%
C	30%	47%

ASHRAE Journal titled "The Facts about Mechanical Grease Filters" (Livchak et. al 2003) presents data on the difference in particulate and total grease removal for various cooking processes and types of mechanical filters. Table 4.4 presents an example from the article of the percentages of particulate and of total grease (particulate plus vapor) removed with three different grease filters when installed over an underfired broiler cooking burgers. Because grease filters installed in commercial cooking hoods are mechanical filters, they only remove the solid grease particles (or grease particulate) from the airstream. Grease vapor, which is in a gaseous state, bypasses the mechanical filters and continues to move through the exhaust ventilation system. For this reason, the total filter efficiency is less than the particulate removal efficiency, because all cooking processes produce both particulate plus vapor. As seen in Table 4.4, there are large variations in commercial kitchen filter performance, which is measured using ASTM F2519 (2019).

Two examples of mechanical grease filters are shown in Figure 4.6. The filter in the top photo is a high-efficiency grease filter, while the filter on the bottom is a standard-efficiency filter. In general, as the pressure drop through the filter increases (either by putting surfaces closer together, changing the airflow path, or adding more collection surfaces) the efficiency increases.

UVC lamps. UVC lamps may be used in commercial exhaust hoods to help mitigate grease levels in the hood plenum and downstream in the ductwork. These lamps produce both UVC light and ozone around the vicinity of the lamps. With UVC lamps, it is important to consider that they must be kept clean to be effective, and to produce ozone, they must be in the temperature range specified by the manufacturer. The lifetime of UVC lamps ranges from 9000–16,000 hours depending on lamp supplier and the specific type of lamp used.

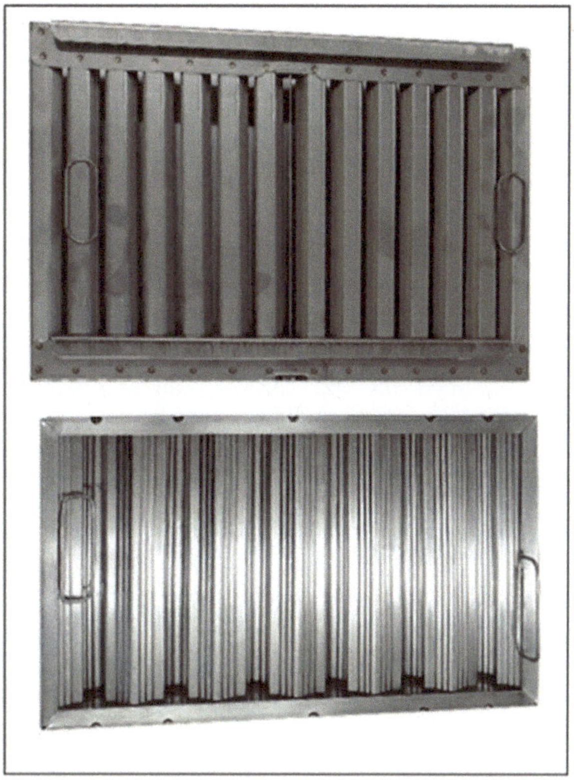

Figure 4.6 Examples of high-efficiency (top) and standard-efficiency (bottom) mechanical grease filters.
Courtesy of Halton Group Americas

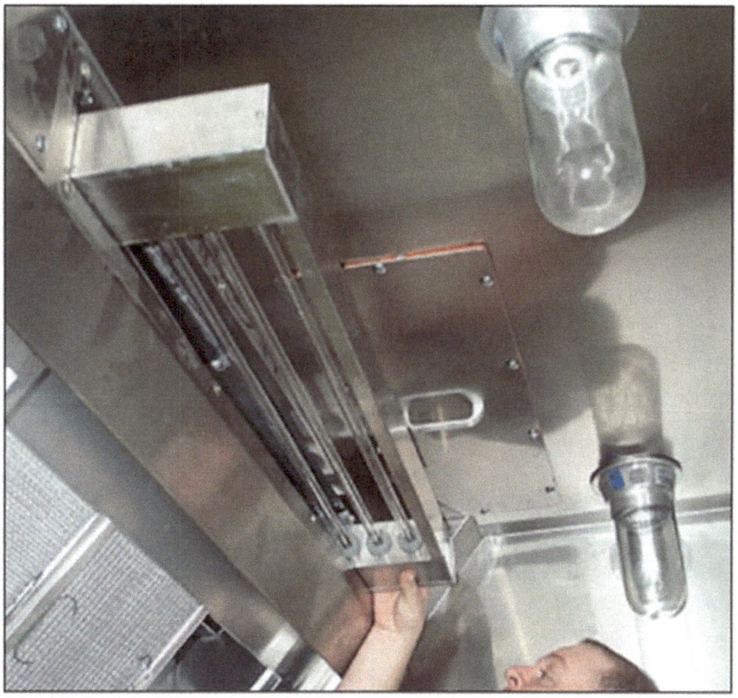

Figure 4.7 Example of UVC lamps in an exhaust hood plenum.
Courtesy of Halton Group Americas

Care should be taken to avoid eye exposure with UVC lamps. Because they produce ozone, they should not operate when the exhaust system is not operational. Figure 4.7 shows a a bank of UVC lamps.

ASHRAE RP-1614 (Zhang et al. 2020), sponsored by ASHRAE Technical Committee (TC) 5.10, Kitchen Ventilation, evaluated UVC lamps in hoods. The results of this study showed that the UVC systems did cause chemical reactions to occur. One conclusion from the study was that the residence time was either too short to see the reactions or that the condensation from grease vapor and grease deposition were small because of limited cooking conducted in the study. Based on those findings, it may be desirable to have a dwell time in the ductwork of 1–2 seconds for ozone to react, depending on the ozone concentration generated with the specific supplier.

Water mist. Water mist is sometimes installed in exhaust hoods over heavy grease and high-temperature or solid-fuel cooking applications. The advantages of mist systems are that they can reduce the exhaust temperatures and help remove or reduce sparks and embers from the airstream. It is important to realize that mist systems tend to operate only when the exhaust fans are turned on and when cooking is occurring.

Water wash. Water wash systems operate during noncooking (idle) periods (commonly at night) to clean the exhaust plenums and sometimes the mechanical filters. These are labor-saving devices that are used in lieu of having staff remove filters to perform manual cleaning of the hood system. An example of a water wash system is shown in Figure 4.8.

ESPs. ESPs are used in some hood systems to help reduce the amount of grease particulate reaching the ductwork. They work by having an ionizer section that charges the grease particulate and collection plates where an electrical field is generated that pushes the particles toward the collection plate. It is important that ESPs are kept clean in order to maintain their efficiency. If too much grease builds up on the ESP cells, these units tend to arc (causing sparking), which causes the efficiency to instantaneously go to zero. It is important that the power supplies used in these applications have an automatic shutdown if sparking occurs. The units are listed to UL 867 (2011b). Figure 4.9 shows a typical ESP cell.

Pollution Control Exhaust Treatment

PCUs are installed downstream of exhaust hoods and may be installed either inline with the ductwork or at the end of the ductwork to help mitigate grease, smoke, and odor emissions. Some of these systems can be the same as previously described in the hood exhaust treatment system. Common components include media filters, ESPs, UVC lamps, carbon filters, HEPA filters, ozone generators, and HEPA filters. For North America, PCUs need to be listed to UL 8782 (2017c).

Media filters. Media filters are the lowest cost grease management device that can be purchased. In a typical filter-based PCU, there are a series of filters installed ranging from low efficiency at the inlet of the unit to higher efficiency after the first filter. These media filters are rated by the minimum efficiency reporting value (MERV) scale in North America and traditionally by a Class scale in Europe. The MERV rating is generated by testing a media filter to ANSI/ASHRAE Standard 52.2 (2017) and the European class is determined by testing either to ISO 16890-1 (2016) for medium or fine filters or to EN 1822-1 (European Standards 2019) and ISO 29463-1 (2017) for efficient particulate air (EPA), HEPA, and ultra-low penetration air (ULPA) grade filters. Table 4.5 compares the MERV and EN 779 and 1822-1 scales. Instead of using MERV ratings, the European standards classify filters as being medium, fine, efficient particulate air filters, high-efficiency particulate air filters, or ultra-low penetration air filters.Note that ISO 29463-1 breaks these classifications into even finer categories.

As these filters become loaded with grease, they need to be replaced with new filters. The loaded filters are typically disposed of in landfills. In PCUs, the typical MERV ratings range from 8 to 15, and these filters are typically staged with less efficient (and less costly) filters being

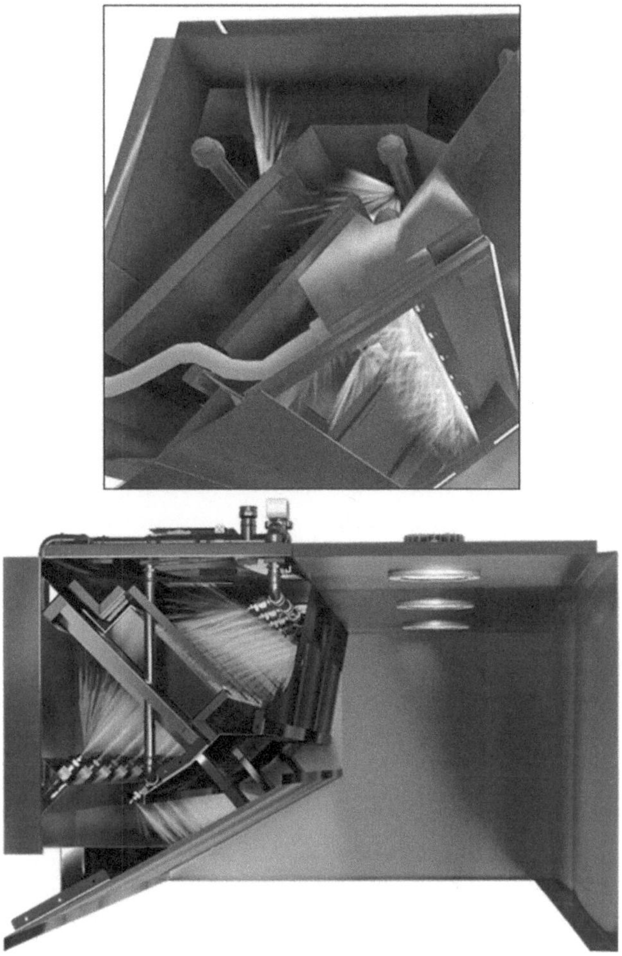

Figure 4.8 Typical water wash in an exhaust hood plenum.
Courtesy of Accurex, a Greenheck Group Company

Figure 4.9 Typical electrostatic precipitator (ESP) cell.
Courtesy of Halton Group Americas

installed before the more efficient ones. This saves cost for the end user, because the less expensive (lower-efficiency) filters collect the larger particulate, and the more expensive (higher-efficiency) filters can be used for a longer period of time before they need to be changed. Figure 4.10 shows a low-efficiency media filter, and Figure 4.11 shows a higher-efficiency bag filter that can hold more grease.

Media filters' lifetimes vary greatly depending on the cooking processes and food product being cooked. Table 4.6 presents the typical life expectancies of various filter categories for medium-duty (i.e., griddle) and heavy-duty (i.e., gas broiler) operations. The table assumes that the filters are used in sequential order, or lowest efficiency to highest efficiency, in a system.

ESPs. ESPs are described in the "Hood-Based Exhaust Treatment" section of this chapter.

MERV 16 and HEPA filters. HEPA filters have a higher efficiency than standard media filters and are typically classified by their penetration efficiency for a particle size of 0.3 microns. A true HEPA filter has a minimum efficiency of 99.95% at 0.3 microns. There are some filters that fall between a MERV 16 media filter and a true HEPA filter in terms of efficiency, as can be observed in Table 4.5. HEPA filters are the best choice when removal of smoke is critical for an application. Figure 4.12 shows a MERV 16 filter. These filters should be installed after lower-efficiency filters.

UVC lamps. UVC lamps are described in the "Hood-Based Exhaust Treatment" section of this chapter. The only differentiator when they are installed in a PCU is that because they are typically closer to the fan termination, the ozone produced by the UVC lamps may not have as much of an impact on the odor as compared to when the lamps are installed in an exhaust hood; this is due to the shorter distances from the lamps in the PCU to the fan or duct termination. The reaction time for ozone created in a PCU might be a fraction of a second versus 1–2 seconds for ozone reactions in the ductwork, depending upon the duct length.

Table 4.5 Comparison of North American (MERV) and European (EN) Filter Ratings

EN 779 (Medium [M] and Fine [F])	EN 1822 EPA (E), HEPA (H), and ULPA (U)	ASHRAE Standard 52.2	Typical Filter Type
M5 ($40\% \le Eff_{0.4} \le 60\%$)	—	MERV 8 to 10	Media
M6 ($60\% \le Eff_{0.4} \le 80\%$)	—	MERV 9 to 13	Media
F7 ($80\% \le Eff_{0.4} \le 90\%$)	—	MERV 13 to 14	Media
F8 ($90\% \le Eff_{0.4} \le 95\%$)	—	MERV 14 to 16	Media
F9 ($95\% \le Eff_{0.4}$)	—	MERV 16	Media
—	E10 ($Eff \ge 85\%$)	MERV 16	Media
—	E11 ($Eff \ge 95\%$)	MERV 16	Media
—	E12 ($Eff \ge 99.5\%$)	—	HEPA
—	H13 ($Eff \ge 99.95\%$)	—	HEPA
—	H14 ($Eff \ge 99.975\%$)	—	HEPA
—	U15 ($Eff \ge 99.9975\%$)	—	ULPA
—	U16 ($Eff \ge 99.99995\%$)	—	ULPA
—	U17 ($Eff \ge 99.999995\%$)	—	ULPA

Figure 4.10 Low-efficiency media filter.
Courtesy of Halton Group Americas

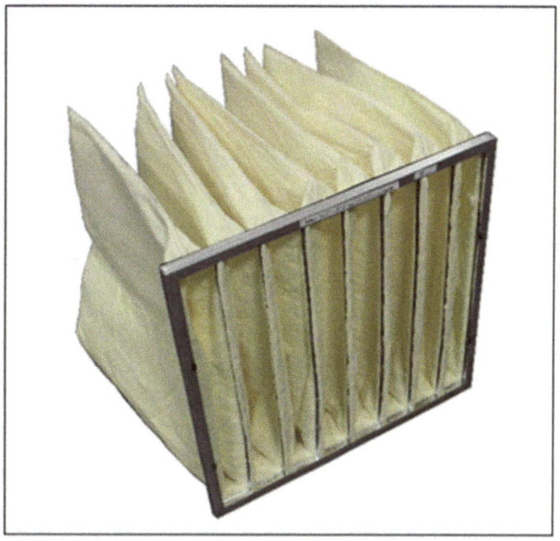

Figure 4.11 Medium- to high-efficiency bag filter.
Courtesy of Halton Group Americas

Table 4.6 Typical Expected Filter Lifetimes for Selected Filters by Cooking Duty

Filter Class	Efficiency	Medium-Duty	Heavy-Duty
Pleated panel filter	M5/MERV 10	2 to 3 months	1 to 2 months
Bag filters	F7–F8/MERV 13–15	4 to 6 months	2 to 3 months
Absolute filter	E11–E12	4 to 6 months	2 to 3 months
HEPA filter	H13–14	2 to 4 months	1 to 2 months

Carbon filters. Carbon filters are commonly used in PCUs to help mitigate odor from the cooking operation. This is an important consideration when designing a system in an urban environment that is close to neighboring buildings or in a mixed-use environment such as being below condominiums. An example of a carbon filter is shown in Figure 4.13. Life spans of carbon filters can vary greatly depending upon the amount of VOCs produced by the cooking process. Generally, a carbon filter's life can range from 2–6 months.

Duct-Based Exhaust Treatment

Ozone generators. Ozone generators, sometimes referred to as *cold plasma systems*, are typically installed in the exhaust system before or after the exhaust hoods and produce ozone while the hoods are operating, which may be beneficial at altering or reducing odors.

TYPE I HOOD FIRE SAFETY

The fire safety systems associated with Type I hoods consist of several aspects, including the following:

- Placement of the exhaust hoods and cooking appliances in relation to building construction materials
- Selection of fire detection and suppression systems for the exhaust hoods
- The regulated interaction between a building fire and the exhaust and supply fans associated with the kitchen
- Appliance/equipment shutoff in the event of a fire
- Any additional requirements specified by the engineer in the sequence of operations

The fire suppression system is designed by the fire suppression supplier and not the design engineer.

Clearance to Combustibles

The reason that clearance to combustibles is an item of concern is that it helps provide a safety margin to allow the fire department time to arrive and extinguish the fire. As a restaurant cooks, the grease builds up in the ductwork over time and may eventually catch fire. It is for this reason that NFPA 96 (2021b) recommends periodic inspection and cleaning of restaurant ductwork. Clearance to combustibles reduces the direct exposure of combustible building materials directly to fire.

NFPA 96 also requires that hoods, PCUs, exhaust fans, ductwork, and cooking appliances have a minimum standard clearance to various types of construction materials as shown in Table 4.7. If a device is listed (per its appropriate UL standard) to be installed with a reduced clearance, that is permitted. *Combustible construction* consists of an assembly where there is a layer of combustible material in the construction (such as having wooden studs covered by gypsum wallboard). *Limited combustible construction* refers to a construction that has a combustible outer material and a noncombustible inner material (such as gypsum wall board on metal studs). *Noncombustible construction* is defined as construction where there is no combustible material (such as masonry walls on metal studs). Note that local codes and standards may have their own requirements for clearance to combustibles that take precedence over national standards.

Hood clearance and appliance clearance. Table 4.7 shows the minimum clearance on the sides, rear, and top of exhaust hoods and cooking appliances. These requirements apply to both unlisted and listed hoods that have not been listed for reduced clearances in accordance with UL 710 (2012). If a hood has been listed for reduced clearances to combustible or limited combustible surfaces, the minimum clearances achieved will vary with the specific products. In a similar manner, the cooking appliances served by the hoods may be listed for reduced clearances to combustibles.

Figure 4.12 MERV filter.
Courtesy of Halton Group Americas

Figure 4.13 Carbon panel-style filter.
Courtesy of Halton Group Americas

Table 4.7 Standard Clearance to Combustibles

Material Type	Minimum Clearance, in. (mm)
Combustible	18 in. (457 mm)
Limited combustible	3 in. (76 mm)
Noncombustible	0 in. (0 mm)

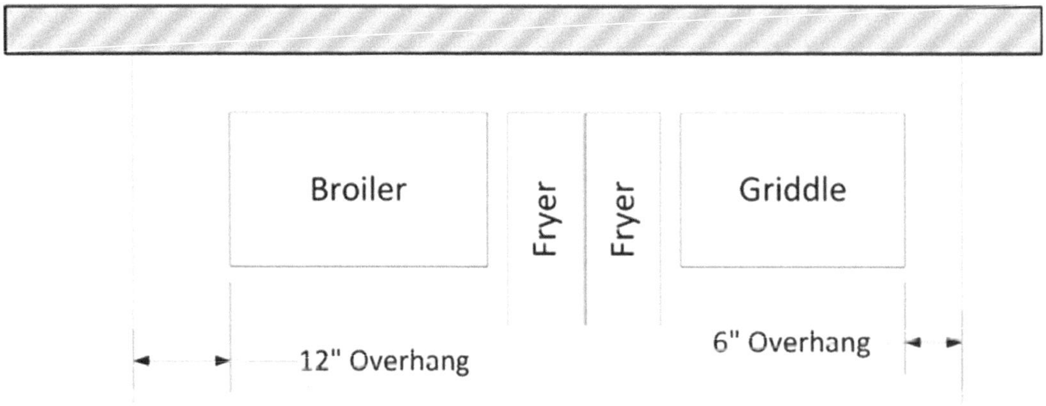

Figure 4.14 Calculating appliance position for a wall-mounted canopy hood.

Hoods and appliances may each have their own clearances that need to be met. For Type I hoods (used where grease is produced), the area of concern is typically the clearance to the rear of the hood (and side of the hood if it is installed in a corner or in an alcove) and the top of the hood. Similarly, for an appliance, there may be different requirements for the side and rear clearances. Additionally, exhaust hoods have an overhang requirement that dictates where the appliance needs to be installed in relation to the side and front of the exhaust hood.

Figure 4.14 shows an example appliance clearance layout, assuming the following:

- The hood is installed adjacent to a wall
- The hood is listed with a minimum side overhang of 6 in. (152.4 mm) for medium-duty appliances and 12 in. (304.8 mm) for heavy-duty appliances
- The hood has an underfired gas broiler on the left side of the appliance line, two deep-fat fryers in the middle, and a gas griddle on the right side of the appliance line

How close could the appliances be to the edge of the hood? To answer this question, calculate how far the appliance would have to be from the side of the hood based on each appliance's duty level. The broiler is considered a heavy-duty appliance and would have a 12 in. (304.8 mm) side overhang, whereas the griddle is considered a medium-duty appliance and would have a 6 in. (152.4 mm) side overhang.

Typically, if an appliance is located in the corner of a room, it can have a 0 in. side clearance to the side panels of a hood or a noncombustible wall surface.

Fire suppression systems. When grease-producing equipment is used underneath a Type I hood (whether that be a ducted or a recirculating hood), it must have an automatic fire suppression system designed for the intended application. (Note that building fire sprinkler systems that are designed per NFPA 13 [2022] are not adequate for Type I hood applications.) The types of fire suppression systems used include carbon dioxide systems, sprinkler systems (which may be water or foam based), dry-chemical systems, and wet-chemical systems (wet-chemical is the most commonly used system). fire suppression systems can be detected through mechanical or electric means. Historically, there have also been water-based fire suppression systems that were covered under NFPA 13 and listed to UL 199 (2022a) for use in hoods or ductwork and UL 199E (2017a) for use over deep-fat fryers. These types of systems are still currently used in some jurisdictions, but they are not used in North American building codes or standards.

Fire protection systems can also be listed in accordance with UL 300 (2019). Figure 4.15 shows the major areas of protection that are required, including the cooking surface of the appliances, the

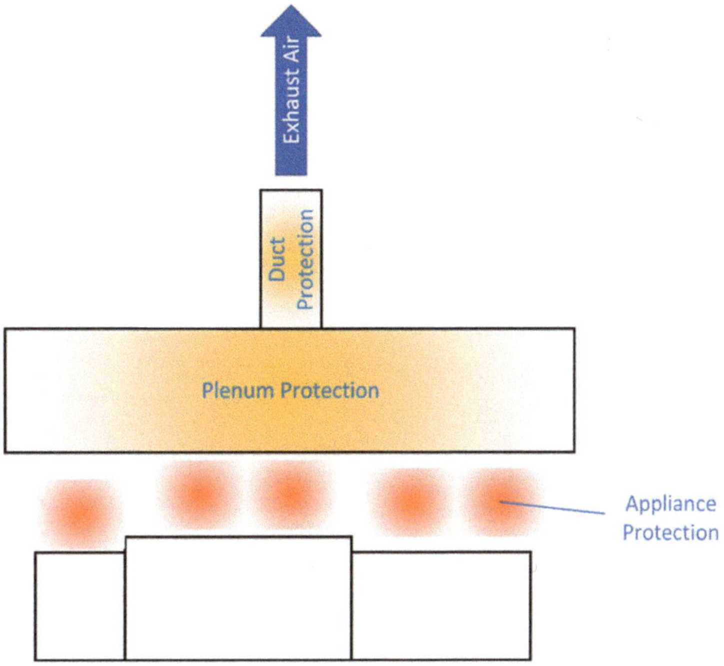

Figure 4.15 Areas where fire protection is required.

exhaust plenum of the hood above the grease filters, and exhaust ductwork exiting the hood. Specific appliances may have customized protection schemes mandated by the AHJ. Figure 4.16 shows a custom fire-protection nozzle for a salamander appliance. (Note that the balloon in the photo is used when the fire suppression system is initially tested with air during commissioning.) Additionally, fire suppression may be required in PCUs.

The fire suppression system may be either an appliance-specific type or a flood type of protection. In appliance-specific coverage, the nozzles are selected for each individual appliance. This type of system is typically less expensive than flood coverage, but if appliances are moved, the coverage needs to be modified accordingly. With flood coverage, the system has more nozzles that are evenly spaced within a hood cavity. This type of system typically has a higher up-front cost but offers the flexibility of being able to move or change the appliances underneath the hood without modifying the suppression system. Some appliances cannot be protected using flood protection; additional piping may be required in those cases.

Mechanical detectors are designed to separate at a specific temperature and release tension on a release mechanism for system actuation. An example of a mechanical detector is shown in Figure 4.17. Electrical detection uses linear or spot heat detectors to send an electrical signal in the event of a fault or fire, allowing the detection line to be continuously monitored. A photo of an electrical detection system is shown in Figure 4.18.

Cooking appliance shutoff. A fire suppression system must automatically shut off all fuel supply and energy to the appliances located underneath that hood or system. Appliances that operate on natural gas typically use a solenoid to close the gas line feeding the appliance. A gas valve may be electrically or mechanically activated via a wall-mounted pull switch or activation of the UL 300-compliant system, which in turn will activate the fire suppression system. All electric appliances may use a shunt trip breaker or some other similar device for shutoff.

Figure 4.16 Custom fire protection for a salamander.
Courtesy of Halton Group Americas

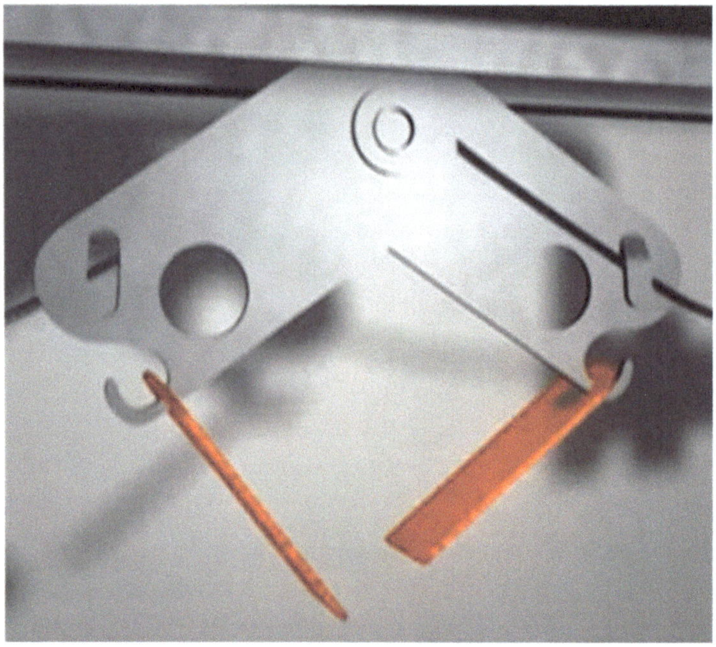

Figure 4.17 Mechanical fire detection system.
Courtesy of Johnson Controls (2022). All rights reserved.

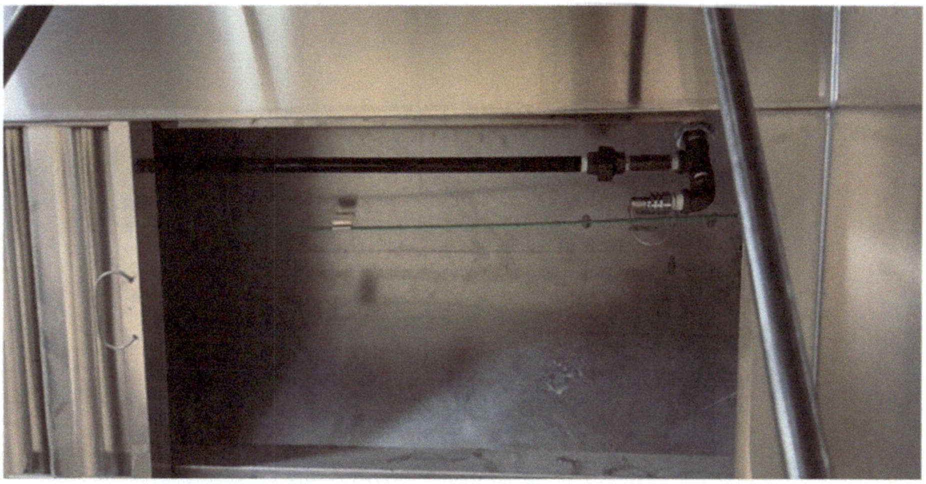

Figure 4.18 Electrical fire detection system.
Courtesy of Accurex, a Greenheck Group Company

Exhaust fan and replacement air operation. In the U.S. and Canada, the typical sequence of operations during a fire condition for exhaust fans and replacement air systems providing air to the kitchen spaces is as follows:

1. The exhaust fan must run at design airflows.
2. The replacement air fan must be shutoff.

These two operations provide the simultaneous benefits of removing smoke from the space (and drawing the fire suppression agent into the hood's exhaust duct), reducing the fresh air (oxygen) actively entering the space, and limiting smoke travel to adjacent spaces.

In some cases, the local AHJ may require that the exhaust fan is shut off or that the replacement air remains on during a fire condition.

Energy Recovery Systems

Energy recovery (also called *heat recovery*) systems are used to reclaim heat from the cooking processes in kitchens that would otherwise be directly exhausted to the outdoors, which conserves energy and reduces the kitchen's operating expenses. One of the most common systems consists of runaround loops that have coils placed in both the exhaust air and supply air to preheat the incoming supply air, as shown in Figure 4.19. In this case, the coils are connected with piping containing a heat transfer fluid. Direct air-to-air heat exchangers cannot be used in commercial kitchens because the exhaust airstream is contaminated with grease and other combustion byproducts; therefore, it cannot be mixed with an incoming outside airstream. Any breaching of the air-to-air heat exchanger would result in the mixing of these two airstreams and cause cross-contamination. Another use of an energy recovery system is to preheat hot water, which can provide a substantial energy savings, depending on the kitchen's operations (e.g., dish washing).

The engineer designing the system usually needs to perform a payback analysis on the system. The systems that are used to preheat supply air are most cost-effective in colder climate zones such as the northern U.S. and Canada, whereas systems that are used to preheat water can be cost-effective in any climate zone. Additionally, use of these types of systems can be a compliance path to

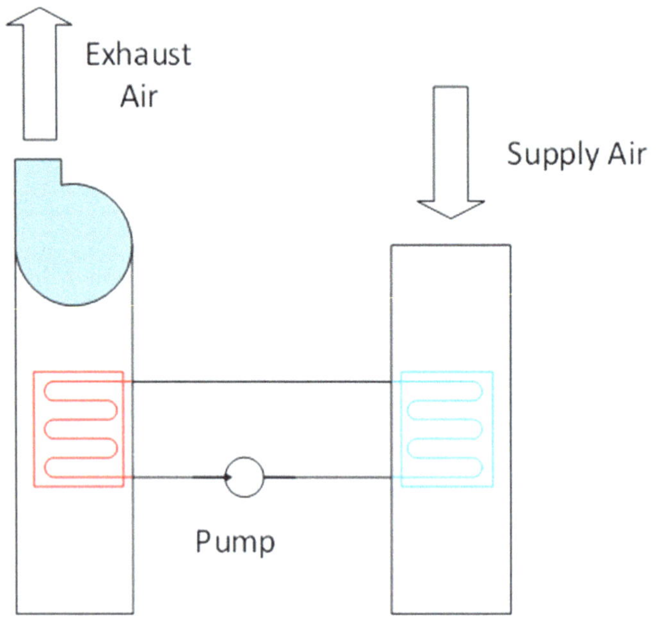

Figure 4.19 Runaround loop to preheat supply air.

meeting ANSI/ASHRAE/IES Standard 90.1 (ASHRAE 2022b), IgCC/189.1 (ICC 2021a), and IECC (ICC 2021c).

Treatment and Bypass

The working fluid in the piping of an energy recovery system is normally treated with an antifreeze, such as glycol, to prevent freezing in the winter months, which is critical if the runaround loop is on the exterior of a building and in a cold climate. Additionally, most systems will have the ability to bypass the coils altogether or shut down the pump when the system is not required. As part of any analysis, the impact of effluent buildup on the exhaust airstream coil must be taken into account. This includes the following:

- Its impact on the coil's heat transfer factors, which will degrade as buildup occurs and will depend on the expected hours of operation and the type of cooking processes.
- The cost of periodically cleaning coils, which will also depend on the type of cooking processes. Note that coil cleaning will likely have to be performed more frequently than typical duct cleaning due to the condensing of effluent on the cool coil surfaces.

While not always required, installing a pollution control device upstream of the exhaust air coil may positively impact the feasibility of the analysis by reducing coil cleaning as well as maintaining a high coil operational heat transfer factor.

A simple payback equation for a runaround loop system recovering heat from the exhaust side and introducing it back to preheat the makeup air during winter is shown as Equation 4.1. Paybacks for this type of system are typically in the range of 10 to 15 years.

$$P = \frac{C}{E_{net}} \tag{4.1}$$

where

P = payback, in years

C = added cost for heat recovery coils, housing, pumps, controls, and piping between exhaust and makeup air units

E_{net} = net annual energy savings calculated as the heating energy replaced by using the heat recovery system minus the additional fan energy required to overcome the additional pressure drop in both the exhaust and makeup air systems (this needs to be determined for each specific geographic location based on the local weather data and the efficiency of the reclaim system)

Costs for a heat recovery system can range from a few thousand dollars to tens of thousands of dollars depending on the size and complexity of the system.

DCKV Systems

Demand-controlled kitchen ventilation (DCKV) systems are probably the most common means of reducing energy consumption in kitchens today (see the "Energy Efficiency Standards" subsection of the "Replacement Air" section of this chapter, specifically regarding ASHRAE/IES 90.1 [ASHRAE 2022] and IgCC/189.1 [2021]). The beauty of these systems is that the exhaust airflow is operated at or near design airflows when cooking is occurring and at lower airflows when cooking is not occurring (the appliances may be either in an idle, ready-to-cook state or they may be powered off). Not only is the exhaust fan energy reduced during these non-cooking and idle times but also, and most impactfully, the energy required to condition the associated heating and cooling of the replacement air is significantly reduced.

These systems have evolved from simple two-speed systems wherein the user manually changed the speed to more complex systems that detect cooking and control variable-frequency drives (VFDs) or electrically commutated (EC) motors to set the exhaust airflow (and corresponding heating and cooling systems' outdoor air quantities) to match precisely with the appliances' operational states (cooking, idle, or off) at any time. Most critically, the exhaust hoods must capture the heat, grease effluent, and smoke when cooking occurs. Regardless of the type of system used, the engineer needs to specify that the system, including all power and control wiring, needs to be wired by the electrician.

Fisher et. al (2013) measured the exhaust fan energy savings for 11 different kitchens using DCKV systems and found that the average exhaust fan energy savings was 41% annually. There are additional savings associated with the heating and cooling systems, dependent on the local climate, but those were not monitored in this study.

What are the most cost-effective applications for DCKV systems? The highest energy savings are typically observed for two cases. The first is when the energy input to the cooking appliances modulates as appliance state varies from off to idle to a full cooking state. A good example of this type of appliance is deep-fat fryers, which consume more energy and therefore produce more of a heat signature during the cooking operation. Conversely, an example of an appliance that may not be a good candidate for DCKV would be an underfired gas broiler or a solid-fuel cooking appliance, both of which have a constant energy input and heat signature regardless of whether the appliance is in an idle or a cooking state. The second case in which the highest energy savings are observed is when there is a lot of diversity with regard to the appliances being off, idle, or in a cooking state. A good example of this is a hotel kitchen where they might only be cooking on one or two appliances during non-peak hours in response to in-room dining orders, and the remainder of the appliances are not being used at that time. (Table 4.10, in the "Energy Efficiency Standards" subsection of the "Replacement Air" section of this chapter, shows where DCKV can be applied as

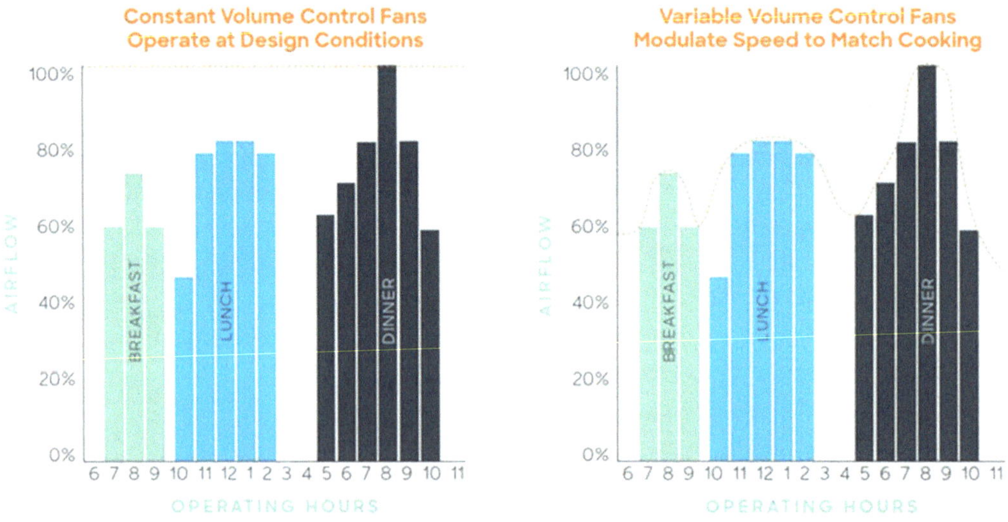

Figure 4.20 Constant-volume airflow versus airflow reduction with a DCKV system.
Courtesy of Accurex, a Greenheck Group Company

an energy savings measure in various codes. It should be noted that the minimum total exhaust airflows vary from 2000 to 5000 cfm [3398 to 8495 cfm].)

It should also be noted that the duct system should be designed to maintain a minimum duct velocity of 500 fpm (2.54 m/s) in accordance with the standards mentioned in this chapter. An example of airflow reduction with a DCKV system, which can lead to energy savings, is shown in Figure 4.20.

The payback of DCKV systems can be calculated using Equation 4.2. Note that in order to calculate the payback, the reduction in airflow needs to be estimated.

$$P = \frac{C}{E_{fan} + E_{heat} + E_{cool}} \tag{4.2}$$

where

P = payback, in years

C = added cost for the DCKV system

E_{fan} = fan energy savings from the reduction in exhaust airflows with the DCKV system along with fan energy reductions in the units supplying the replacement air to the space (for an exhaust fan, the power [hp or kW] can be estimated using the fan law, shown in Equation 4.3)

E_{heat} = annual energy savings during the heating season based on the reduction in replacement air airflows; typically equal to the quantity of exhaust air reduced

E_{cool} = annual energy savings during the cooling season based on the reduction in replacement air airflows; typically equal to the quantity of exhaust air reduced

$$P_2 = P_1 \cdot \left(\frac{n_2}{n_1}\right)^3 \tag{4.3}$$

where

P_2 = calculated power of fan, hp (kW)

P_1 = power of fan at design airflow conditions, hp (kW)

n_2 = fan speed at reduced airflow, rpm

n_1 = fan speed at design airflow conditions, rpm

Types of Cooking Sensors

DCKV systems also use a broad spectrum of technologies to determine what state (cooking, idle, or off) the appliances are in, from duct- or canopy-based temperature sensors to cooking activity sensors (such as optical detection or infrared sensors). One important aspect of system selection is how quickly the hoods go to full design airflow when cooking occurs. Schrock et. al (2012) performed research on various types of detection systems and concluded that the systems with cooking activity sensors had a much faster response time than those systems using only temperature sensors. Additionally, they found that in several cases the temperature-sensor only systems failed to achieve full design airflow during cooking, which means those hoods may not have been capturing all of the heat, smoke, grease, and other effluents of cooking.

Another important selection criterion for DCKV systems is how low the airflow can be reduced below design airflows when the appliances are in idle operation. Fisher et. al state that they "believe that the potential turndown (average cfm reduction) for the more sophisticated DCKV systems can be greater than the temperature-only systems" (2013, 49).

Hood and Fan Combinations

One of the decision points when designing a CKV system is the number of fans to use relative to the number of exhaust hoods. Factors that should be considered include the project's cost and whether exhaust fan redundancy is required.

Often the simplest type of system to design and control is when a single hood is connected to a single exhaust fan. In this case, from a DCKV perspective, it is straightforward to know how each fan needs to operate based on the cooking appliances underneath the corresponding hood. An additional benefit to this approach is that it creates redundancy, so if one fan fails, the other fans/hoods continue to operate. Ductwork runs can be simple for low-rise buildings but become more problematic for high-rise facilities. A potential negative with this approach is that there are additional costs for the separate roof penetration, fan, and electrical services for each hood along with added maintenance for each of the fans.

Another system design involves connecting multiple hoods to a single exhaust fan, commonly referred to as a *manifold* design. This approach can be especially beneficial where long duct runs are required and the space for ductwork is limited. From a DCKV perspective, all of the hoods have to run at full design airflow if any single appliance underneath any of the hoods is cooking. The addition of continuously variable balancing dampers that can regulate the airflow on each hood according to the appliance operation can control the air exhausted from each hood. In this type of system, there are additional costs for these balancing dampers along with the added maintenance. The other potential challenge is that there is no redundancy for a given exhaust system—if the sole exhaust fan fails, all associated exhaust hoods are effectively shut down.

In a case study that had four hoods connected to a single exhaust fan, Schrock et. al (2012) determined that the additional use of balancing dampers doubled the energy savings when compared to the same system without balancing dampers installed. It should be noted that the same energy savings can be achieved if each hood has its own dedicated exhaust fan.

Control of Makeup Air Systems

The IMC (ICC 2021d) specifies that makeup air (or replacement air) be supplied to the kitchen whenever the exhaust hoods are operating and that the kitchen space maintain a slightly negative pressure relative to adjacent spaces. The space balance for the kitchen should be evaluated at the full range of exhaust airflows. When DCKV systems are employed on the exhaust hoods, the most common means of doing this when communicating with makeup air units (MAUs) is to send a signal to the MAU VFD or EC motor to vary the speed of the makeup air to match the air being exhausted through the hood. Control systems might also consist of using various configurations of rooftop units (RTUs) and dedicated outdoor air system (DOAS) units to properly balance the makeup air into the kitchen. For larger buildings, it is also common to have the exhaust airflow from the DCKV system sent to the main building maintenance system (BMS) so the system can control the regulation of the makeup air.

It is important to understand that the turndown capability of the system as a whole is dependent on the turndown capability of the exhaust and supply fans. Additionally, some MAUs may be limited in their amount of airflow turndown. This is true of direct-fired gas burners, which require a certain pressure drop (or air velocity) across them to provide heating. If the airflow in the unit falls below the lower pressure limit during heating, the combustion will be shut off and unheated air would be provided at that and lower airflows. Indirect-fired gas heaters typically have the ability to operate at a much higher turndown ratio than direct-fired units.

DCKV Commissioning

All DCKV systems should have commissioning performed by an approved manufacturer representative and a qualified third party (commissioning provider) to ensure proper performance at all operating points. To achieve proper performance, the commissioning provider should, prior to commissioning, follow all manufacturer installation guidelines and ensure that manufacturer drawings and literature are referenced during the installation. The installation typically involves low-voltage control and communication wiring between a network of hoods and control panels. Additionally, if VFDs are used to modulate exhaust and makeup air, they must be installed and configured correctly prior to any commissioning procedure.

To properly commission a DCKV system, a range of operating conditions should be evaluated. Because most systems include a manual override function that will drive exhaust hoods to design airflow, this is a good place to begin the process, as it will also verify that the air balance is per the design drawings. All HVAC systems and equipment associated with the DCKV must be operational and have been properly balanced prior to this step. With the manufacturer's assistance, the commissioning agent will begin by manually putting all hoods in minimum airflow (idle) conditions to ensure that the makeup air modulates accordingly to properly pressurize and balance the space. Simulation of multiple cooking operating points is recommended next. For example, if hood operation is staggered based on meal periods, the commissioning agent will simulate these conditions to ensure proper kitchen-space balance. During this process, the commissioning agent will ensure that all connectivity between any BMS is verified. In some instances, control signals will come from multiple sources; these must be verified to work in harmony for proper operation.

Once equipment is verified to be operating correctly, full-load cooking is recommended to ensure proper capture and containment (C&C) performance. The commissioning agent will coordinate with the site owner to perform actual cooking processes, which typically can only occur after the certificate of occupancy has been issued by the AHJ (so this testing will have to be scheduled accordingly). Partial-load cooking should also take place to verify proper system operation. In many instances, a manufacturer's representative will assist with the process and make adjustments to control components to optimize performance.

Many DCKV systems offer customization options to integrate with the BMS or cooking appliances. Ensure that any custom features or functions are documented and are able to be tested during the Commissioning Process.

Retrocommissioning is strongly recommended to ensure proper operation of the DCKV system through its operable lifetime. When changes to a kitchen footprint or process occur, retrocommissioning should also be performed. See chapter 8, "TAB and System Commissioning," for additional information.

Exhaust Ductwork

Exhaust ducts for commercial kitchens are classified into two categories in the national building codes: Type I and Type II. Type I ductwork is for use when serving a Type I exhaust hood, which is used when grease is produced by the cooking process; Type II ductwork is used when serving a Type II hood, which is used when only heat and steam are produced (such as with dishwashers). See chapter 5, "Grease Exhaust Duct and Kitchen Hoods.

Type I Ductwork

The overarching purpose of Type I ductwork is to manage the cooking effluent that gets through the exhaust hood and provide a safe system in case of fire. As previously described, the hood or PCUs remove some—but not all—of the contaminants at various locations in the system. Additionally, these ducts serve the purpose of assisting with smoke evacuation in case of a fire in the building.

Duct velocity. For systems that operate at a constant exhaust airflow (i.e., non-DCKV systems), the duct velocity is typically in the range of 1500 to 1800 fpm (7.6 to 9.1 m/s). Refer to section 5.3 in ANSI/ASHRAE Standard 154 (2022d) for more information. When systems are designed with a DCKV system, the ducts are typically designed for a duct velocity of 500 fpm (2.5 m/s) at the lowest exhaust airflows. Currently that valve is the lowest duct velocity allowed by NFPA 96 (2021b), IMC (ICC 2021d), and ASHRAE Standard 154 (2022d). Note that the listed 500 fpm minimum duct velocity is due to the minimum velocity that was tested as part of ASHRAE RP-1033 (Kuehn et al. 2000). It was the researcher's opinion that, based on their observations, a lower duct velocity would still provide safe operation, but the test equipment was not capable of properly operating at a lower duct velocity, so the 500 fpm (2.5 m/s) minimum was established.

Construction and grease duct clean outs. Type I ductwork must be constructed and installed so that it is liquid-tight, which means that it does not allow pressurized water (e.g., from a pressure washer) or grease to leak out of the duct. This may be accomplished either by applying a liquid-tight weld or brazing at the joints in the field or by using a duct system that has been listed to show that it meets these conditions. Additionally, clean outs must be provided so that the ductwork can be regularly cleaned to reduce fire risk. Refer to appendix B, "Research Conducted by ASHRAE TC 5.10," for more detailed information on clean outs and duct construction.

Fire-rated enclosures. When fire-rated enclosures are required, the ducts may be installed in a fire-rated enclosure constructed around the ductwork, listed field-applied grease duct enclosures can be used, or ductwork that is listed with an integral fire-rated enclosure may be used. Refer to appendix B, "Research Conducted by ASHRAE TC 5.10," for more detailed information on fire-rated enclosures.

Exhaust outlet termination. Per the IMC (ICC 2021d), Type I exhaust outlets (including grease exhaust fans) must terminate not less than 40 in. (1.02 m) above the roofline and not less than 10 ft (3 m) horizontally from parts of the building, property line, or adjacent buildings. The outlet termination also must be not less than 10 ft (3 m) horizontally or not less than 3 ft (0.9 m) vertically above air intake openings.

Ductwork is allowed to terminate through an exterior wall where the smoke, grease, gases, vapors, and odors do not create a public nuisance. It is a good design criterion to also account for the direction of prevailing winds and other building features that may cause exhaust air to be reentrained into the building. For more detailed information, refer to IMC Section 506.3.13, "Exhaust outlets serving Type I hoods" (ICC 2021d). Table 4.9 of this chapter contains expanded information on the termination requirements for exhaust fans.

Solid-fuel cooking equipment. Per the IMC (ICC 2021d), ductwork serving a Type I hood that is used over solid-fuel cooking equipment must have its own duct, exhaust fan, and fire suppression system. Solid fuel cooking equipment is not allowed to share a hood with other non-solid-fuel Type I appliances that require automatic fire suppression. For a detailed overview of all code requirements for solid fuel cooking equipment refer to NFPA 96, chapter 15, "Solid Fuel Cooking Operations" (2021b).

Type II Ductwork

Type II ductwork has fewer requirements than Type I ductwork, because it is used where no grease is produced. The primary construction criterion is that it be constructed of rigid metallic materials that are not easily corroded by water vapor (e.g., aluminum or stainless steel). Type II exhaust ductwork is not allowed to be incorporated with Type I exhaust ductwork.

Exhaust outlet termination. Type II exhaust outlets must not terminate less than 3 ft (0.9 m) from a building opening, less than 10 ft (3 m) from the property line or other buildings, less than 10 ft (3 m) above ground, less than 30 in. (762 mm) above a roofline, or less than 30 in. (762 mm) from a vertical wall. For more detailed information, refer to the IMC, Section 506.4.2, "Type II terminations" (ICC 2021d).

Exhaust Fans

Exhaust fans are the devices that do the work to convey the air through the exhaust hoods and ductwork to the building exterior. As with exhaust hoods and ductwork, exhaust fans are differentiated by whether they are used with Type I (grease-laden) hoods or Type II (heat and water) hoods.

For exhaust fans used with Type I hoods and ductwork, the fans must be listed to UL 705, Supplement 5C (2017b), for the U.S. and ULC-S645 for Canada (UL 2021b). UL 705 contains requirements previously in UL 762 (2013), *UL LLC Outline of Investigation for Power Roof Ventilators for Restaurant Exhaust Appliances*. These fans should contain and properly drain all grease removed from the airstream, and the motor needs to be protected from the grease-laden airstream.

When exhaust fans are used with Type II hoods and ductwork, it is not required but it is recommended that they be listed to UL 705 (2017b) for the U.S. and CAN/CSA C22.2 No.113-18 (CSA Group 2018) for Canada. This is recommended to protect the motors from excess moisture and heat.

Sizing Fans and Motors

The power of the fan required to move air is a function of the airflow, static pressure, and fan efficiency, as shown in Equation 4.4. Equation 4.5 shows the calculation to determine the motor power to operate the fan. Note that for direct-drive fans, the drive efficiency is 1.0, and for belt-drive fans, it is typically 0.97.

$$P_F = 0.000157 \cdot Q_s \cdot p/\eta_F \tag{4.4}$$

where

Q_s = flow rate at standard conditions, cfm (cmh)

P = static pressure, in. of water (Pa)

P_F = power required at fan shaft, hp (kW0

η_F = fan efficiency, dimensionless

$$P_M = (1 + DL) \cdot P_F/(E_M \cdot E_D) \tag{4.5}$$

where

P_M = power required at input to motor, hp (kW)

DL = drive loss, dimensionless

E_M = fan motor efficiency, dimensionless

E_D = drive efficiency, dimensionless

Static Pressure

When sizing a commercial kitchen exhaust fan, there are several sources of static pressure that the fan needs to overcome, including losses at the exhaust hood filtration and collars, ductwork, PCUs, and, if the fan is inline, the losses after the fan until it terminates to the outdoors. Figure 4.21 shows typical sources of static losses. The first source is the exhaust hood, which consists of losses at the filter (1) and the exhaust collar (2); these components are typically provided by the hood manufacturer. The duct losses (3) need to be determined by the mechanical engineer on the job, and the losses of other systems (4), such as PCUs, need to come from the respective manufacturers. All these losses are summed together to determine the total static pressure for which the exhaust fan needs to be designed. For manifolded duct systems (where multiple ducts are connected to a single duct), the hood and duct system with the largest pressure drop are used when sizing the exhaust fan.

Fan Types

The most common fan types used for Type I applications include power roof ventilators (PRVs), known as *upblast fans* (Figure 4.22); centrifugal fans, also known as *utility set fans* (Figure 4.23); and in-line fans (Figure 4.24). PRVs can also be mounted in a sidewall configuration (Figure 4.25) when routing ductwork through the building proves challenging for a particular

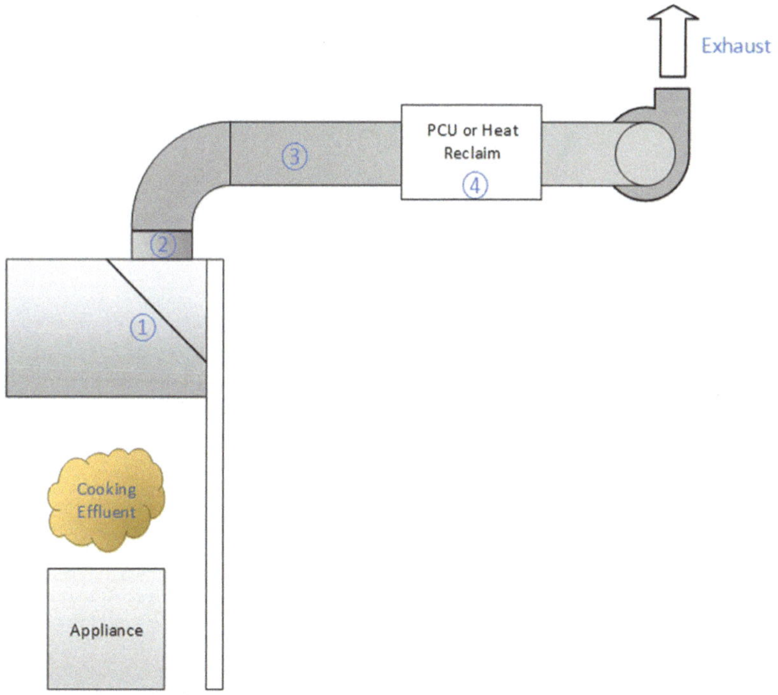

Figure 4.21 Typical static losses in a commercial kitchen exhaust system.

Figure 4.22 Upblast fan.

Courtesy of Halton Group Americas

Figure 4.23 Utility set fan.

Courtesy of Halton Group Americas

Figure 4.24 In-line fan.
Courtesy of Accurex, a Greenheck Group Company

Figure 4.25 Sidewall-mounted exhaust fan.
Courtesy of Accurex, a Greenheck Group Company

Table 4.8 Fan Applications

Consideration	Upblast	Utility Set	In-line	Sidewall
Reduced first cost	x			x
Ease of cleaning	x			x
Indoor mounting		x	x	
Change direction of airflow		x		

Table 4.9 Grease Fan Requirements

Requirement	Upblast	Utility	Inline	Sidewall
Height to fan base from roof deck (above grade for sidewall)	18 in. (457 mm)	18 in. (457 mm) if mounted outdoors	N/A	10 ft (3 m) unless in secured area or allowed per local AHJ
Height to fan outlet from roof deck	40 in. (1.02 m)	40 in. (1.02 m) if mounted outdoors	N/A	N/A
Horizontal clearance from outlet to adjacent buildings, property lines, air intakes	10 ft (3.05 m)	10 ft (3.05 m) if mounted outdoors	N/A	10 ft (3 m)
Horizontal clearance from outlet to any combustible structure	5 ft (1.52 m)	5 ft (1.52 m) if mounted outdoors	N/A	
Vertical clearance from outlet of fan (above) to air intakes within 10 ft (3 m) of exhaust outlet	3 ft (914 mm)	3 ft (914 mm) if mounted outdoors	N/A	10 ft (3 m) (includes doors and windows)
Other requirements	Hinged, grease trap, flexible weatherproof electrical cable, flanged, and gasketed fan base	Grease trap; flanged, gasketed, and bolted duct-to-fan connections	Where enclosed duct systems are connected to in-line fans not located outdoors, the fan must be located in a room or space having the same fire-resistance rating as the duct enclosure. Access must be provided for servicing and cleaning of fan components. Such rooms or spaces must be ventilated in accordance with the fan manufacturer's installation instructions.	All ductwork must be pitched to drain the grease back into the hood(s) or have a drain provided to bring the grease back into a container within the building or into a remote grease trap. The exhaust flow must be directed perpendicularly outward from the wall face or upward.

application. Table 4.8 shows the optimal conditions under which to use an upblast fan, a utility set fan, or an in-line fan based on field considerations and first-cost considerations. Table 4.9 shows applicable code requirements for each type.

Where entrainment of the effluent back into the replacement air systems is a concern, high-velocity or extended discharges may be used. Refer to *ASHRAE Handbook—HVAC Applications* (2019), chapter 34, "Kitchen Ventilation," for more information on the high-plume fan for this application.

System Effects

System effects occur when conditions at the fan inlet and outlet leads impact the distribution of air, thereby negatively impacting the system's performance. Fans are tested to produce accurate performance data, including the static pressure internal to the fan as a system. Part of this testing criteria includes inlet and outlet conditions that lend to a uniform velocity profile (see Figure 4.26) entering and exiting the fan system. Conditions that can greatly impact fan performance include elbows too close to the inlet and/or outlet of the fan, abrupt duct transitions near the fan, inlet spin due to duct design, dampers and their locations, poorly designed guards, and anything that could block or affect the trajectory of air coming into or leaving the fan (see Figure 4.27 for examples of good and poor conditions). If such installation conditions cannot be avoided, it is important that

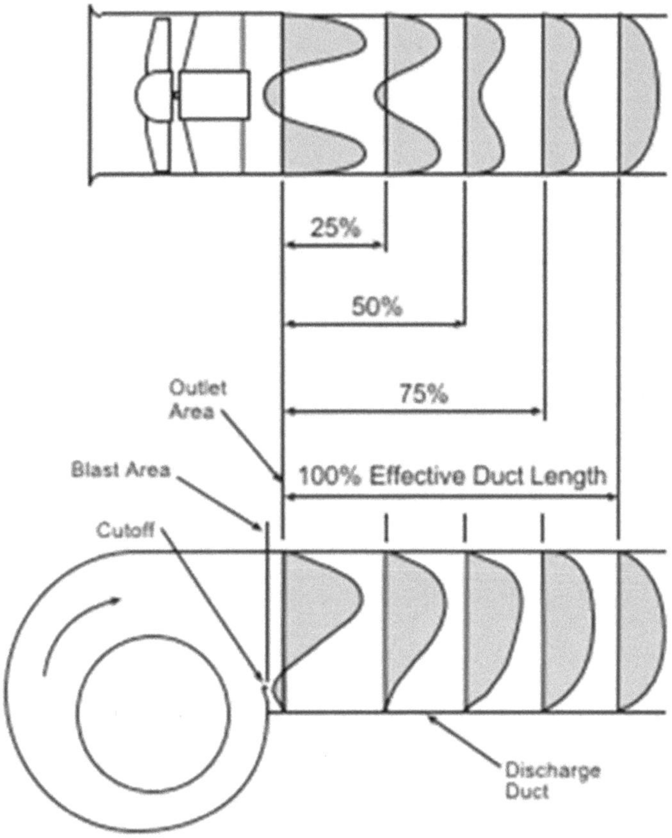

Figure 4.26 Two fan examples showing the effective duct length at which the velocity is uniform.
Courtesy of Accurex, a Greenheck Group Company

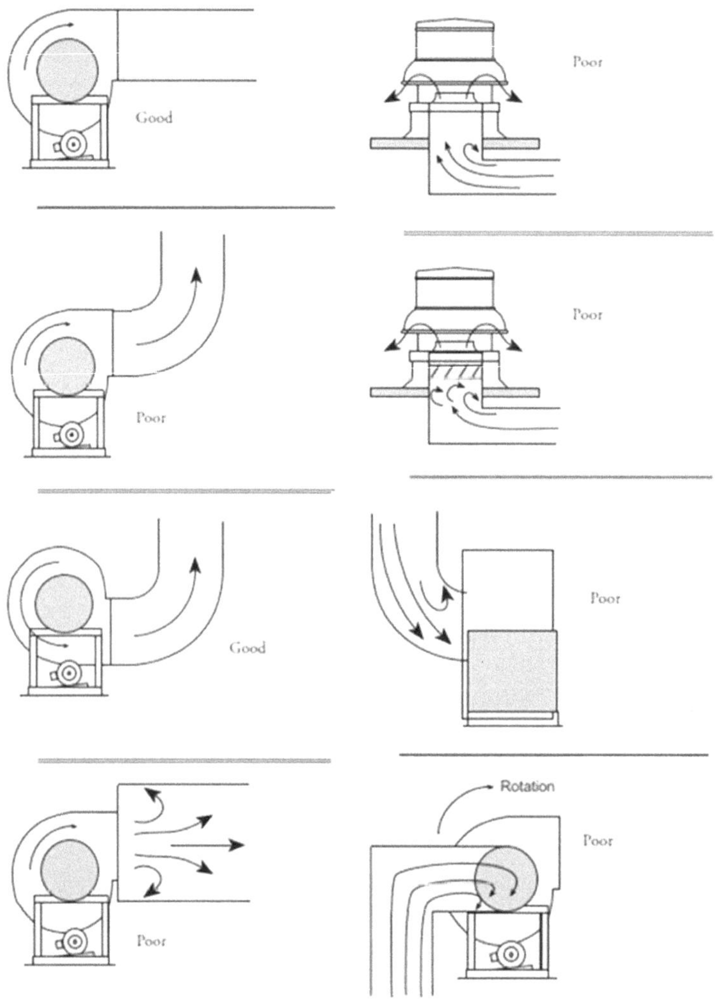

Figure 4.27 **Good and poor inlet and outlet conditions for installed fans.**
Courtesy of Accurex, a Greenheck Group Company

they be taken into consideration when selecting a fan by estimating the impact of the system effect on performance and adding that into the system's resistance calculation during design.

Replacement Air

When the air exhausted from the hoods is discharged outdoors, that air must be replaced in the kitchen with other sources of air. Those sources include transfer air, supply air, makeup air, and, undesirably, infiltration or exfiltration air. Infiltration refers to the air brought into the kitchen or restaurant through cracks, openings, or events such as door openings; exfiltration refers to the air exiting those same spaces. Figure 4.28 depicts these common replacement air sources. This section also describes the heating and cooling and dehumidification equipment used to provide replacement air.

Makeup Air

MAUs (also known as *dedicated replacement air units*) are a specific type of replacement air device commonly used in commercial kitchens to replace the air that is being exhausted through

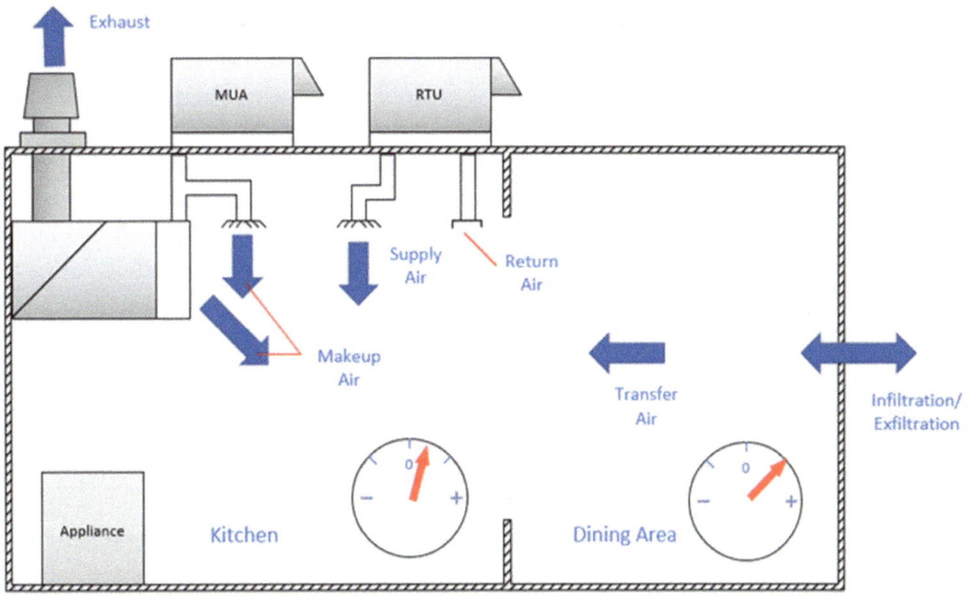

Figure 4.28 **Replacement air methods for commercial kitchens (outdoors is neutral or zero pressure).**

the ventilation hoods. This section compares and contrasts them to traditional RTUs and DOAS units, which are designed with the goal of ensuring that the design temperature and humidity in a commercial kitchen space are maintained by handling not only the heating and cooling loads from the outdoor air but also all the internal loads from cooking equipment (hooded and unhooded), ventless hoods, people, solar gain on the building shell, infiltration/exfiltration, and other ancillary equipment in the space. RTUs and DOAS units are discussed in more detail below. It is also important to note that the use of transfer air from a dining area to the kitchen can be preferred over the use of makeup air as a method of bringing in replacement air, because the air has already been conditioned to a comfortable space temperature with relatively low humidity levels.

Design Considerations

MAUs are different than RTUs and DOAS units in that they are designed not to handle the internal heat gains but to provide unconditioned (also known as *untempered*) or partially conditioned (also known as *partially tempered*) outdoor air to supply to the space that does not negatively impact the comfort or humidity in the commercial kitchen space. In moderate climates, untempered air may be used in the summer or winter months to provide replacement air to the kitchen space.

The best design practice when providing cooling from an MAU is to provide a supply air temperature of no more than 4°F (2.2°C) above the design temperature kitchen when it is located in a humid- or mixed-weather zone or within no more than 8°F (4.4°C) above the design temperature when it is located in a dry- or arid-weather zone. When heating, it is recommended that the supply air be not less than 8°F (4.4°C) below the space design temperature. Additionally, if colder or hotter air is supplied near the hood, it can negatively impact hood performance. Some building codes, such as the IMC, state that "the temperature differential between the makeup air and air in the conditioned space shall not exceed 10°F (5.6°C) except where the added heating and cooling load do not exceed the capacity of the HVAC system" (ICC 2021d, section 508.1.1). The following items are associated with this approach:

- If there are significant moisture loads being brought into the building through the MAU, the walls and other building structures may start absorbing moisture, which can lead to mold growth. The HVAC system may not have the capacity to cool and dry the surfaces after they become wet.

- Most kitchen HVAC systems are not designed to reflect the actual loads to the space (including the heat and potential moisture gains from the makeup air systems).
- There is a common myth that all of the air from MAUs introduced in plenums near the exhaust hood is immediately exhausted directly from the hoods and does not cause load in the space. Chapter 34, "Kitchen Ventilation," of *ASHRAE Handbook—HVAC Applications* (2019) recommends that engineers assume that at least 50% of the outdoor air heat and moisture that are introduced into the kitchen space are accounted for as a heat load to the kitchen space.

The general components of an MAU include an intake hood and filter section, a heating section, a cooling section, and a blower. The exact order and configuration of these components varies from manufacturer to manufacturer, but a general representation is shown in Figure 4.29.

Typical installation for MAUs in commercial kitchens is either on the roof, in which case the unit typically has a downward vertical discharge to connect to supply ductwork, or inline with the ductwork, in which case it typically has a horizontal discharge out of the end of the unit. The intake or rain hood for the units (when mounted outdoors) typically has mesh filters to prevent larger debris (such as leaves) from entering the unit. Most manufacturers also offer dust filters that can be installed in or after the intake hood.

Supply Air

Supply air that is conditioned is brought in through either a packaged RTU or a DOAS. The real differentiator between these two systems is that RTUs are typically limited to bringing in approximately 25% of outdoor air during cooling, while DOAS units are designed to supply 100% conditioned outdoor air. If an RTU is used in economizer mode, it may be able to bring in up to 100% outdoor air when there are favorable outdoor air conditions.

Transfer Air

Transfer air reuses air that is brought into the dining space to meet the occupancy ventilation requirement and transfers that air into the kitchen to replace part of what is being exhausted through the hoods. This air is fully conditioned and complies with ANSI/ASHRAE Standard 62.1

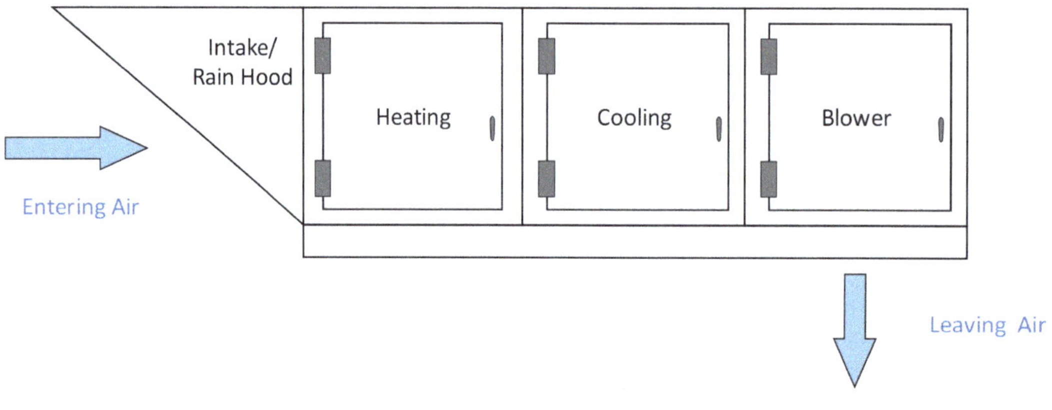

Figure 4.29 Schematic of a typical MAU.

(2022c). For this case, note that the outdoor air is treated in the dining space and then transferred to the kitchen.

Heating Systems

The most common types of heating systems include direct-fired gas heat, indirect-fired gas heat, electric heat, and hot water coils.

Direct-fired gas units. Direct-fired gas units may use natural gas or propane and typically consist of a gas burner that is placed directly in the unit's airstream, as shown in Figure 4.30. These gas burners may be installed horizontally or vertically, depending on the unit design.

The advantage of these units is that they have a 100% combustion efficiency with a thermal efficiency of 92%, but the drawback is that they are putting some combustion byproducts (carbon monoxide and nitrous oxide) into the space. However, in spaces with many air changes (such as kitchens), these byproducts are less of a concern. Direct-fired systems are tested to ensure they operate only within a safe range of combustion byproducts. It is recommended that combustion byproduct sensor alarms be included when using direct-fired gas units.

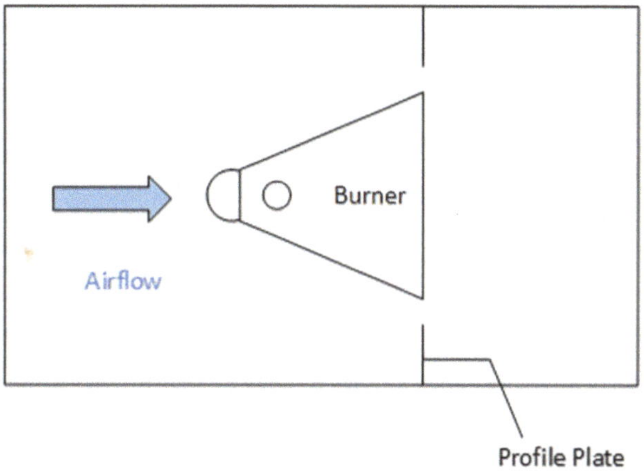

Figure 4.30 Typical direct-fired gas burner.

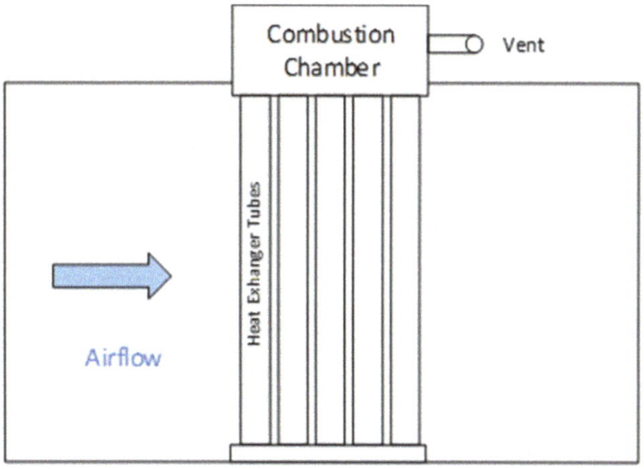

Figure 4.31 Typical indirect-fired gas heater.

Indirect-fired gas units. Indirect-fired units use a furnace located outside of the airstream to preheat incoming air with a heat exchanger, as shown in Figure 4.31. Indirect-fired units have a lower efficiency (maximum 80%) compared to direct-fired units, and they tend to be more expensive, larger, and more difficult to maintain than direct-fired units. While the combustion byproducts are emitted outside the building, care needs to be taken so that they do not get reentrained into the building through one of the supply air inlets. These units must also be capable of operating at the required outdoor air design temperatures without the combustion gases condensing in the heat exchanger, which can cause corrosion and premature heat exchanger failure.

The advantage of indirect-fired gas units is that they do not directly introduce combustion byproducts into the space. Care needs to be taken to ensure that the vented combustion byproducts are not reentrained into any nearby building supply air systems, because they are highly concentrated and toxic. As stated previously, the other disadvantages of indirect-fired gas units are that they have a lower combustion efficiency (maximum 80%), are larger in size, and are more expensive than direct-fired gas burners to achieve the same heat output.

Electric. These systems use electric resistance elements in the airstream to preheat the air. This type of system is easy to operate and does not have the safety concerns associated with gas-fired systems, but the electric resistance option has equivocally higher operating costs due to the peak demand utility charges and greater utility costs than the gas-fuel equivalent, and they may require a larger building electrical service, especially in colder climates.

Steam and water coils. Steam or hot-water coils are a good choice for heating the incoming air where district steam or excess hot water is available. The advantage is that they are a low-maintenance option, but the engineer must also design the hydronic loop, including the pumps, freeze protection (if applicable), and control systems.

Heat pumps. When gas is not available on site, another option for heating is to use a heat pump. The primary advantage of a heat pump over electric resistance is that it is more efficient. The heating performance for heat pumps is stated as a heating seasonal performance factor (HSPF). The required minimum values vary depending on the type of unit and capacity. ASHRAE/IES Standard 90.1 (ASHRAE 2022b) lists minimum required HSPF values. Care also needs to be taken that heat pumps are not used in too-cold climate zones where the conditions are beyond the unit's design capabilities and the condensing coils therefore may freeze in the winter. Also, some heat pumps may use electric resistance heat as a secondary heat source, which can lead to higher operating costs.

Cooling and Dehumidification Systems

The primary types of cooling systems used in replacement air systems include direct expansion (DX) cooling and evaporative cooling. Evaporative cooling systems only perform well in hot, dry climates, because they use water to help cool the incoming air but they do not provide significant dehumidification. DOAS units specifically use hot-gas reheat to control the humidity, which provides finer temperature and humidity control than DX coil-based systems that do not have hot-gas reheat. A chilled-water coil could also be installed to provide cooling; those can be integrated with chiller loops or heat pump systems. Another means of controlling humidity is to use desiccant dehumidification systems, although they are not very common in commercial kitchen applications, because the grease from any nearby exhaust fans can enter into the makeup air intake, become impregnated into the desiccant material, and reduce its ability to operate.

DX cooling systems. For climate zones that have humid- or mixed-weather conditions during the summer months, DX units are the most commonly used systems. These systems generally use R-410A as the refrigerant, may have evaporator coils inside the MAU to cool the air, and may use either a package (condenser is integrated into the same unit) or a split system design where the condenser is mounted remotely from the evaporator. The advantage of a package system is that it comes prepiped and charged from the factory, whereas with a split system, field installers need to run the connecting copper piping between the evaporator and condensing units and charge the units in the field. The advantage of split systems is that they tend to be available in larger capacities than package units and are more flexible in terms of roof placement. With a DX system, the unit can have the capacity to reduce both the sensible and latent loads of the incoming air.

One drawback of DX cooling systems is that the units only dehumidify while the condensers are operating, and if the units are oversized, they have a tendency to quickly cycle on and off, which can lead to large variations in the humidity levels in the space (it will increase when the condensers are off and decrease when they are operating).

Evaporative cooling. For climate zones that are dry (have low humidity), evaporative cooling systems are the most cost-effective option. These systems have a pad that is wetted with water, and as hot outdoor air is drawn across the pad, the water on the pad evaporates and cools the air. The advantages of this type of system are their simplicity and efficiency compared to a DX cooling system. Additionally, these types of systems can be used where mechanical cooling systems are not permitted. Disadvantages of this system include that they have lower performance during rainy periods (when incoming humidity increases) and they have increased water consumption compared to a DX system.

Chilled-water coils. When a facility has chilled water available to use, chilled-water coils can provide a low-maintenance means of cooling. As with hot-water coils, the engineer needs to design the piping, pump, and control system for this type of configuration.

Integration with DCKV. Most MAUs can be ordered in either a constant-volume or variable-air volume (VAV) configuration. Building codes require that replacement systems, including makeup air systems, be interlocked with the exhaust ventilation system so that as the exhaust airflow changes, the supply airflow also changes to maintain proper space pressurization. VAV systems—and their associated VAV terminal units—can be easily configured to operate in conjunction with a DCKV system. Kitchen spaces typically need to maintain a pressure of –0.01 to –0.02 in. w.c. (–2.5 to –5 Pa) relative to the other spaces in the building. This is partly done to keep cooking odors in the kitchen. VAV systems are most commonly controlled using either a VFD or an EC motor. Generally, belt-drive fans use a VFD, while direct-drive fans may use either a VFD or an EC motor. EC motors are becoming more common and offer cost and efficiency advantages. EC motors are commonly offered in sizes up to 2 hp (1.5 kW) in single-phase configuration and up to 15 hp (11 kW) in three-phase configuration.

With MAUs, attention needs to be paid to how low the airflow can be turned down during heating and cooling seasons. In general, most direct-fired gas MAUs allow a maximum turndown of at least 2 to 1, or 50%, on the airflow during heating mode, which means that they can provide heating down to half of the design airflow. Some manufacturers have proprietary burner or damper configurations that may allow an even higher turndown. Most motors have a maximum turndown ratio of approximately 7 to 1, or down to 30%, of the design revolutions per minute.

Impact on Hood Performance

The most important aspect of replacement air is designing it such that it does not negatively impact the hood performance during normal operation. If the replacement air has too high of a velocity at the front of the hood, or if it can reach the thermal plume from the appliances, it may cause the hood to spill, which means that combustion byproducts, grease, smoke, heat, and other contaminants could spill to the space. Research on commercial kitchen makeup air found that in order to maintain the hood performance, the velocity of replacement air reaching the front of the hood should not exceed 75 fpm (0.38 m/s), although lower velocity is even better (CEC 2002). Regardless of the type of replacement air used, the air is brought into the kitchen space through various styles of plenums, grilles, or diffusers, or it can be integrated into the hood itself. The following subsections present the benefits and challenges of each style; additional information can also be found in the results of research projects, such as the Brohard et al. (2003) report on makeup air effects on commercial kitchen exhaust system performance.

When designing a kitchen space, the replacement air diffusers for bringing fully conditioned air into the space should ideally be located as far away from the hood as possible. This is so that the replacement air can help provide comfort and clean air to the entire space and minimize negative impacts on the hood performance. It is also recommended to install these diffusers in locations where they do not discharge air onto any unhooded food-processing appliances or holding areas, because the cool air will also cool down cooked food.

Perforated ceiling diffusers. There are two broad types of perforated ceiling diffusers: the first is supply plenums located along the exposed sides of a hood (commonly referred to as *perforated perimeter supply*), and the second is diffusers that are located in the ceiling but not adjacent to the hood. For perforated perimeter supply, the air velocity may need to be reduced so it is within the recommended velocity range when the air reaches the front lip of the exhaust hoods.

Perforated celling diffusers that are not adjacent to the hood need to be evaluated so that their throw pattern does not hit the front of the hood or the plume from the cooking operation. In commercial kitchens, these types of diffusers are much better than directional diffusers.

Directional ceiling diffusers. Two- and three-way directional diffusers should be installed so that their air is not discharged toward any hoods. The disadvantage of these types of diffusers is that the contractor can easily install them in the wrong orientation, which will negatively impact hood performance. Four-way ceiling diffusers, if used, should not be installed within 15 ft (4.6 m) of any hoods. Generally speaking, four-way diffusers are not recommended for use in commercial kitchens.

Displacement diffusers. This style of diffuser operates at a low velocity, making it ideal from a hood-performance perspective. The challenge in kitchens may be finding enough ceiling or floor space to accommodate the diffusers.

Front-face (hood) supply air. Front-face supply air can work well and has the advantage of reducing the ceiling space needed to supply the air relative to other diffusers located in the ceiling. This style of diffuser, if properly designed, directs air away from the front of the hood and can operate at higher discharge velocities, up to 150 fpm (0.76 m/s).

Slot diffusers and nozzles. Slot diffusers and nozzles tend to have an extremely high discharge velocity and are not recommended for use in commercial kitchens.

Short-circuit (internally compensating) air. Short-circuit (internally compensating) air is makeup air that is introduced internally into the hood. In practice, it has been found that in order to maintain the required net exhaust airflows, the exhaust airflow needs to be increased to maintain the hood performance. Therefore, this type of makeup air introduction is not recommended in commercial kitchens. ASHRAE/IES Standard 90.1 (ASHRAE 2022b) limits the amount of internal makeup air to no more than 10% of the exhaust airflow.

Back-wall supply. Back-wall supply systems bring makeup air behind and below the line of appliances. While these types of systems do not negatively impact hood performance, they are not recommended for providing unconditioned makeup air, because makeup air that is colder or hotter than the kitchen design temperature can negatively impact thermal comfort when the makeup air is colder or hotter than the design temperature in the kitchen (Livchak et al. 2005). If the air is conditioned, the better design approach is to introduce it into the kitchen space and provide comfort to the occupants.

Energy Efficiency Standards

ASHRAE/IES Standard 90.1 (ASHRAE 2022b), IgCC/189.1 (2021), and the 2022 *California Building Standards Code* (CEC 2022), commonly known as Title 24, all have energy efficiency

Table 4.10 Energy Efficiency Standards as They Relate to CKV

Energy Efficiency Measures	90.1	IgCC/189.1	Title 24
Applies when kitchen has an airflow greater than X	X = 5000 cfm (2360 L/s)	X = 2000 cfm (950 L/s)	X = 5000 cfm (2360 L/s)
Transfer air	At least 50% of replacement air	At least 50% of replacement air	At least 50% of replacement air
DCKV	Installed on 75% of exhaust air	Installed on 75% of exhaust air	Installed on 75% of exhaust air
Energy recovery devices	An energy recovery ratio not less than 40% on at least 50% of the exhaust airflow	An energy recovery ratio not less than 40% on at least 50% of the exhaust airflow	An energy recovery ratio not less than 40% on at least 50% of the exhaust airflow
At least 75% of makeup air must met these criteria	Not applicable	Not applicable	• Be unheated or heated to no more than 60°F (15.6°C) • Be uncooled or cooled without using mechanical cooling

components that require some form of energy efficiency system be used to save energy, as shown in Table 4.10. All of these standards allow the use of transfer air as replacement air in the kitchen space, allow some form of DCKV system, and allow the use of sensible heat recovery to capture and reuse what would otherwise be wasted heat energy. Additionally, these standards specify a maximum hood exhaust airflow depending on the type of hood and duty level of the cooking appliances underneath the hood. Title 24 also has a fourth option that allows for unconditioned makeup air. Outside of California, it is not recommended to use unconditioned makeup air, especially in areas that are hot in the summer and/or cold in the winter.

5

Grease Exhaust Duct and Kitchen Hoods

There are two types of kitchen hoods, Type I and Type II. This chapter discusses exhaust ducts for the various styles of Type I hoods including duct construction, installation, and maintenance thereof to remove the grease-laden vapors and effluent created by the cooking process.

It should be noted that exhaust ducts for Type II hoods are constructed of rigid metallic materials that are not easily corroded by water vapor (e.g., stainless steel and aluminum). If the Type II exhaust operates with a positive static pressure or contains moisture-laden air, the duct must be constructed, installed, and sealed accordingly. Single-wall ducts (modular factory built or field fabricated) are often used for Type II hood exhaust applications; however, double-wall insulated ducts can also be used if clearance to combustibles or skin temperatures are a concern. Refer to chapters 3 and 4 for more information on Type II hoods.

Duct Design General Principles

Exhaust duct routings should be as short as possible to the terminal outlet. This creates less static pressure and less potential for grease accumulation. It also leads to long-term operational savings for the building owner due to the reduced time needed for maintenance and system cleaning. The Type I hood grease duct system should not be interconnected with any other building ventilation or exhaust system.

Common industry design grease duct velocities are from 1500 to 1800 fpm (7.6 to 9.1 m/s) at design airflows (ASHRAE 2019). National mechanical codes, design handbooks, and standards allow for a minimum of 500 fpm (2.5 m/s). Duct noise and/or excessive static pressure should be considered if significantly exceeding the industry design velocity ranges, and maximum velocities should not exceed 2500 fpm (12.7 m/s) (ASHRAE 2019). If multiple hoods combine into a single common exhaust duct run, each junction fitting (i.e., a tee) and downstream duct should increase in size to maintain the preferred design duct velocity ranges. The duct designer or engineer of record should be fully aware of the mechanical code having jurisdiction for the exhaust duct system, as there are some differences between the requirements of various codes (e.g., NFPA 96 [2021b] versus *International Mechanical Code* [IMC, ICC 2021d]).

Construction

Duct Materials

There are two types of materials used for grease exhaust ducts. One option is factory-built stainless steel grease ducts (the inner wall is required to be stainless steel) that are *listed* and labeled in accordance with UL 1978 (2010d) test standards and permitted to use materials in accordance with their listing. Listed grease duct manufacturers can use lighter gauge materials due to their UL testing and listing (~0.035 in. [0.89 mm], no. 20 MSG). They can be round duct profile but can also be oval or rectangle/square.

The other option is to field-fabricate the duct by minimum code construction methods that are *unlisted*. Code requires unlisted fabricated ducts to be constructed of carbon steel not less than 0.0575 in. (1.46 mm, no. 16 MSG) in thickness or stainless steel not less than 0.045 in. (1.14 mm, no. 18 MSG) in thickness. The shape profile is usually a rectangle or square.

Duct Joint Connection and Duct Testing

There are typically two types of duct joint connection methods. One type is a UL 1978 (2010d) listed, non-welded joint connection that is mechanically locked and sealed and follows the factory-built grease duct's installation instructions. The other type is an unlisted field-fabricated duct joint that requires a liquid-tight, continuous, external weld on all seams, joints, penetrations, and duct-to-hood collar connections.

Mechanical code requires that the duct joints be evaluated after installation for leakage; there are two methods of evaluation. One method is the traditional 100 W light bulb test, which is applied to duct joints, seams, and hood connections (exceptions apply to listed ducts). The 100 W light test is applied to all field welds to see if light passes through to identify any holes/pinholes in the welds. For listed ducts, the light test is only to be applied to field-assembled joints. ANSI/ASHRAE Standard 154 (2022d) does not include the light bulb test, because it has not been shown to be a reliable leak test method in the field.

The other suggested method, per ASHRAE Standard 154 (2022d) and NFPA 96 (2021b), is a water wash-down test using a high-pressure spray gun that would reveal leaks on any improper joint or fabrication methods for listed or unlisted duct. This is often more practical, as it can be applied to the entire duct system from the high end of the slope back toward the hood and replicates future cleaning operations that the duct system will have to endure. This initial wash-down testing also prompts the initial construction stakeholders (installers, cleaners, inspectors) to open access doors and find post-installation "leftover" items, such as light bulbs or cords left in ductwork and electrical wires otherwise potentially not discovered until first cleaning (see Figure 5.1).

(a) (b)

Figure 5.1 (a) Light bulb and cord and (b) and electrical wires left in ductwork and not discovered until first cleaning.

Courtesy of DPW Services

Duct Support Spacing

Duct supports and guides are made of noncombustible construction. Listed factory-built (UL 1978) ducts follow their support/load and spacing distances relative to their listing and installation manual. They have greater loads and distances than unlisted field-fabricated ducts. UL 1978 (2010d) requires a fourfold safety factor be applied to the installation manual's published loads and distances.

Unlisted field-fabricated ducts usually follow the Sheet Metal and Air Conditioning Contractors' National Association (SMACNA) guidelines (2020) requiring supports that are 60 in. (1525 mm) on center for horizontal ducts and every floor vertically. If personnel entry into the ductwork for maintenance and cleaning is planned for, the unlisted field-fabricated duct supports must be designed to accommodate the additional weight of an occupant (NFPA 96 [2021b]; IMC [ICC 2021d).

Installation

Clearances to Combustibles (Not Within a Fire-Rated Enclosure)

Single-wall ducts (unlisted field fabricated or listed factory built) require 18 in. (457 mm) clearance to combustibles. Listed factory-built duct manufacturers also offer several options for double-wall ducts that have reduced clearances ranging from 0 in. to less than 18 in. (457 mm), and if 0 in. clearance is not necessary (no adjacent combustible building materials), the non-zero clearance double-wall models are more economical and still provide similar overall safety and performance benefits.

For unlisted field-fabricated single-wall ducts, zero clearance can be accommodated by installing two layers of a field-applied listed grease duct wrap.

NFPA 96 (2021b) and IMC (ICC 2021d) also specify clearances to noncombustible construction and limited-combustible construction (or gypsum wall board attached to noncombustible structures). Reference those sources accordingly based on job-site jurisdiction.

Enclosures

In general, when a grease duct passes through an hourly fire-rated barrier, it must have a vertical or horizontal enclosure around it of the same fire rating (≤ 1 hour), and the enclosure usually must be continuous to the building exit. Note that there is an exception where an enclosure is not required: when a grease duct penetrates only a non-fire-resistance-rated roof/ceiling assembly.

There are currently three methods of enclosure construction, as outlined below:

- **Traditional rated shaft construction (per the *International Building Code* [ICC 2021b]).** Clearances are 18 in. (457 mm) for single-wall ducts within a combustible shaft and 6 in. (152 mm) for single-wall ducts within a noncombustible shaft (includes gypsum wall board attached to noncombustible construction or an NFPA limited-combustible definition). Clearances for listed, non-fire-rated, double-wall factory-built grease ducts are allowed for reduced clearances per their listing. Ducts are sealed at the point of entry and ventilated to the exterior through weather-protected openings.
- **Factory-built enclosure assemblies listed and labeled to UL 2221 (2010e) with corresponding firestop system in accordance with UL 1479 (2015) or ASTM E814-13a (2017).** These should be installed according to the manufacturer's instructions.
- **Field-applied enclosure assemblies listed and labeled in accordance with ASTM E2336 (2020a) with corresponding firestop system listed in accordance with ASTM E814-13a (2017).** These are flexible grease duct wraps applied to minimum-code-constructed unlisted field fabricated grease ducts. This flexible wrap is required to be applied from the hood con-

nection to the building exit and to cover/wrap the duct on all sides. Mechanical codes do not permit partial flexible wrapping of a duct segment for the sole purpose of clearance reduction in a specific location. Ducts that are greater than 24 in. (610 mm) wide require steel banding and weld pinning (bottom side) or weld pinning on all four sides. Any exposed foil-faced flexible wraps subject to such potential contact should be protected from physical damage.

A duct enclosure serves a single grease duct and does not contain other ducts, piping, or wiring systems unless acceptable to local jurisdiction.

NFPA 96 (2021b) has similar duct-enclosure language and some fire-rating criteria that is relative to the building height as follows:

- Less than four stories requires an enclosure hourly rating of not less than one hour. Exception: Single-story buildings without a requirement for an hourly-rated roof/ceiling assembly do not require an enclosure.
- Four stories or more requires an enclosure hourly rating of not less than two hours.

There are other restrictions such that fire dampers and smoke dampers are not permitted within grease ducts. Exceptions do apply for bleed air ducts. Refer to NFPA 96 (2021b) for complete details.

Grease Duct Slope

Horizontal grease duct systems require slope (pitch) to mitigate stagnant grease effluent accumulation, and the duct construction should never form dips or traps within its layout. The slope usually flows back toward the hood or sometimes, but not often, toward an in-line grease drain (NFPA) or reservoir (IMC) that needs to be fully accessible and maintained. Drains or reservoirs are included when proper slope of the system is not achievable, as these components help guarantee proper removal of grease buildup. With the use of grease drains or reservoirs, the building owner should be made aware that regular observation and maintenance of these drainage points is required, possibly separate from their usual cleaning intervals.

A factory-built listed grease duct system can be installed to minimum slope requirement per the manufacturer's listing to UL 1978 (2010d). Typically, for round and oval listed grease ducts, the slope is as low as 1/16 in/ft (5.2 mm/m) for horizontal runs that are less than 75 ft (22.9 m) in length. For horizontal straight runs that are greater than 75 ft (22.9 m) in length, a minimum 3/16 in/ft (15.6 mm/m) is required.

Field-fabricated flat-bottom unlisted ductwork requires a minimum slope of 1/4 in/ft (21 mm/m) for horizontal straight runs that are less than 75 ft (22.9 m) in length, and a minimum of 1 in/ft (83.3 mm/m) is required for runs that are greater than 75 ft (22.9 m).

A "seesaw" duct design approach (i.e., peaks and valleys) has been acceptable in some jurisdictions for job-site conditions with very minimal overhead vertical space. Systems following this design are required to include grease drains or reservoirs at the low end of each valley (see Figure 5.2).

Many factory-built listed grease duct manufacturers allow for a hybrid duct system layout using listed factory-built and unlisted field-fabricated and field-wrapped (i.e., rectangle duct) to accommodate low overhead vertical space. It is important for the duct designer to work closely with the duct manufacturer to guarantee the most efficient design is planned.

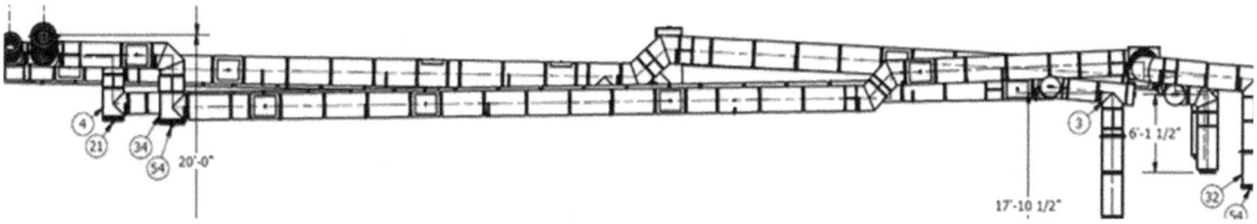

Figure 5.2 "Seesaw" (up/down) duct run layout when minimal vertical overhead space is available.
Courtesy of Selkirk Corp/Hart & Cooley (c/o Duravent Group)

Access Door Openings

Well-planned access to a grease duct system is imperative to fire protection and the safety of the building and its occupants. Proper access allows for the necessary cleaning measures to occur, eliminating the accumulation of flammable grease over time. The IMC (ICC 2021d) and NFPA 96 (2021b) require the following:

- A minimum of one access door opening for personnel entry on horizontal ducts of at least 20 × 22 in. (508 × 559 mm) (IMC) or 20 × 20 in. (508 × 508 mm) (NFPA).
- Access doors must be spaced at a minimum of 20 ft (6.1 m) (IMC) or 12 ft (3.7 m) (NFPA) apart for horizontal ductwork if personnel entry is not possible.
- An access door is to be located within 10 ft (3 m) of a change in direction that is greater than 45° (IMC) or at every change in direction (NFPA).
- Access door openings are not required in duct sections that are accessible from the duct entry (from the hood) or exhaust termination discharge. Access must be provided at the top of the vertical riser to accommodate descent where personnel entry is possible. Per NFPA requirements, if the duct size is too small for personnel entry (per the previous statement), an access door is required on every floor level.

IMC (2021) section 506.3.8, Item 1, states, "Grease ducts shall not have openings except where required for the operation and maintenance of the system," which infers that designers should minimize openings in a duct system to keep the system as practical as possible.

The designer should be aware that not all doors spaced according to code language can be physically used once the ductwork and all of the other building trade systems are in place. If a duct layout cannot be improved due to building architectural limitations, a practical code variation petition to an authority having jurisdiction (AHJ) should be considered in lieu of having any amount of inaccessible doors installed. Examples of code variations are a reduction in door usage and/or a change in location spacing distances. It should be understood that an exhaust duct system with fewer doors (that can still be reasonably accessed) becomes a safer, more economical, and more energy-efficient system than a system with doors installed per code that are inaccessible. Having doors located and appropriately sized for a future duct cleaner can help optimize a safe and thoroughly maintained system. Also, consider involving a local credentialed/certified duct cleaner in the initial duct design layout for any given project.

Figure 5.3 shows examples of unreasonably located and completely inaccessible grease duct access doors.

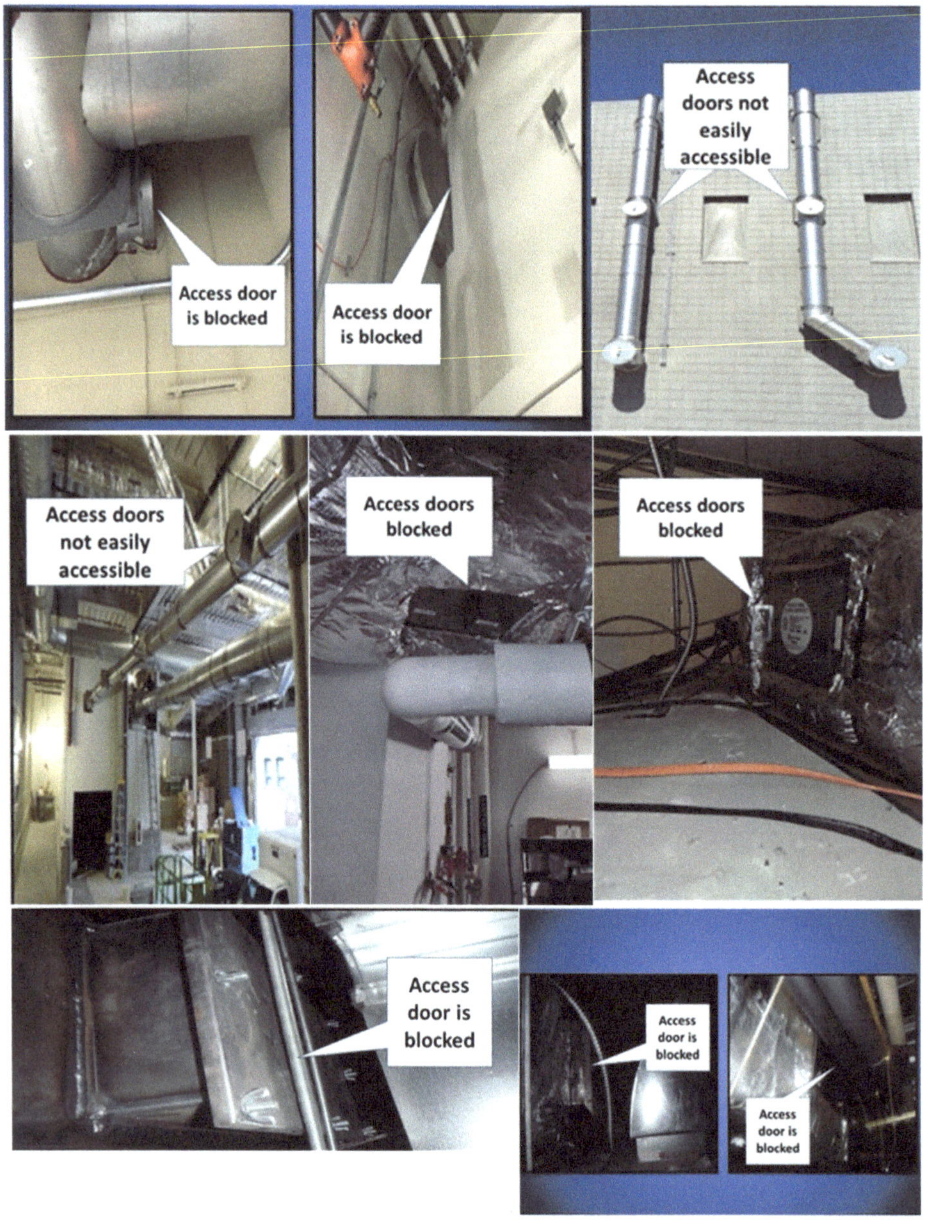

Figure 5.3 Unreasonably located and completely inaccessible grease duct access doors.
Courtesy of DPW Services

Exhaust Outlets Serving Type I Hoods (Terminations and Fan Connections)

There are usually two common termination types for grease duct: rooftop vertical discharge or (sidewall) horizontal discharge. All exhaust fans should be appropriately listed to service grease effluent.

Rooftop vertical discharge terminations are usually accomplished by means of an upblast fan mounted onto a roof curb. Another option is for the duct to pass through the roof deck (through a thimble, flashing, or curb) and go horizontally into a utility-set fan. A less-common exhaust approach is to have an in-line fan within the building interior. In-line fans also create a positive pressure downstream within the duct work. This is not the most ideal situation compared to having the fan at the end of the system pulling a negative on the exhaust stream, because positive-pressure ductwork downstream of the fan has the potential to cause leaks into the building.

Rooftop terminations can also be accommodated by having the ductwork go through an exterior wall and up to the roof and 90° elbow back over onto the roof deck into a utility-set fan. When ductwork will be exposed to the atmosphere in colder climates, it is highly recommended that a double-wall insulated listed grease duct be used to help mitigate grease effluent cooling and solidification.

Sidewall horizontal discharge terminations are usually accomplished by a wall-mounted fan, which is similar to an upblast fan except that it is horizontally mounted. Sidewall terminations can also be accomplished by in-line fans, but the caution relating to the positive pressure inside the ductwork applies. Note that NFPA 96 (2021b) restricts wall terminations to only noncombustible wall construction.

The codes require termination/discharge clearances required for vertical and horizontal terminations, so it is important that the designer fully understand them as they relate to adjacent building surfaces, openings, or walkways. See chapter 4 of this book for more information. Fan vibration and dampening should also be considered for utility-set and in-line fan applications.

Solid-Fuel Cooking

NFPA 96 (2021b) includes much detail related to solid-fuel cooking, so that standard should be referenced in these applications.

Solid-fuel cooking exhaust must be separate from all other exhaust systems, with the exception of solid-fuel/smoke flavoring applied to gas or electric cooking operations.

Wall terminations are not permitted for solid-fuel cooking exhaust systems.

Duct systems that are three stories or less (with or without a hood over the cooking operation) require a grease exhaust duct of either listed factory-built or unlisted field-welded construction (per code requirements). For duct systems that are four stories or more, solid-fuel cooking requires the duct to be a listed factory-built duct design.

Some solid-fuel cooking appliances (i.e., pizza ovens) allow for direct connection of the ductwork (or chimney) per listings.

In general, most listed round stainless steel factory-built grease ducts are also multi-listed as all-fuel chimneys (UL 103 [2010a]) and therefore designed to withstand the internal fire risks associated with both grease and creosote deposition. Hotter exhaust temperatures help mitigate the formation of creosote, so consideration should be made for direct connection to solid-fuel-burning appliances (if a direct-duct connection is available) and also shorter duct routes with limited exposure outside of the building's thermal envelope.

Fire Endurance Testing

Factory-built round, stainless, duct tested/listed to UL 1978 (incl UL 2221) has repeatedly shown that it inherently resists profile deformation during the extreme UL internal grease fire test (2000°F [1093°C] for 30 minutes). Comparatively, it has been demonstrated that unlisted, field-welded, 16 gauge rectangular steel duct loses its shape profile (i.e., collapses) during this same 2000°F (1093°C) test, requiring major demolition/rework thereafter. Some other notes of interest are provided below:

- UL grease duct standard testing restricts a certain amount of reduction in cross-sectional area after the severe internal grease fire testing of 2000°F (1093°C) for 30 minutes.

- ASTM E2336 (2020a) standard testing, which is what most flexible grease duct insulation wraps (with foil-face covering) are listed to, does not have a maximum reduction in cross-sectional area and therefore permits the unlisted field-welded duct to fully collapse down during testing and still qualify for passing the test.
- Listed factory-built grease ducts require consistent quality fabrication per inspections by the listing agency, whereas listed flexible grease duct insulation wraps are only inspected by their listing agency for the insulation wrap performance attributes and not any actual ductwork.

Severe grease buildup (see Figure 5.4), inaccessible duct access doors, and routine cleaning negligence are very likely to exist, therefore increasing the risk of internal grease fires within the duct.

Figure 5.5 shows UL grease duct internal testing that includes ramping up to 2000°F (1093°C) for approximately 30 minutes to replicate a severe internal grease fire. This test is performed immediately after an ~4-hour 500°F (260°C) operation test.

Figure 5.6 shows a listed round stainless duct before and after the UL severe internal test at 2000°F (1093°C) for 30 minutes.

Figure 5.7 shows an unlisted wrapped rectangular carbon steel duct before, during, and after a UL severe internal test.

Grease Duct Cleaning

The construction of listed round stainless steel ducts has better long-term corrosion resistance than that of unlisted carbon steel ducts against duct-cleaning detergents. Over time, this can play a large role in mitigating potential leaks into the building space. Listed round stainless steel ducts usually make the cleaning process easier and quicker because of the round shape profile and lower sur-

Figure 5.4 Grease buildup in rectangular unlisted duct.
Courtesy of DPW Services

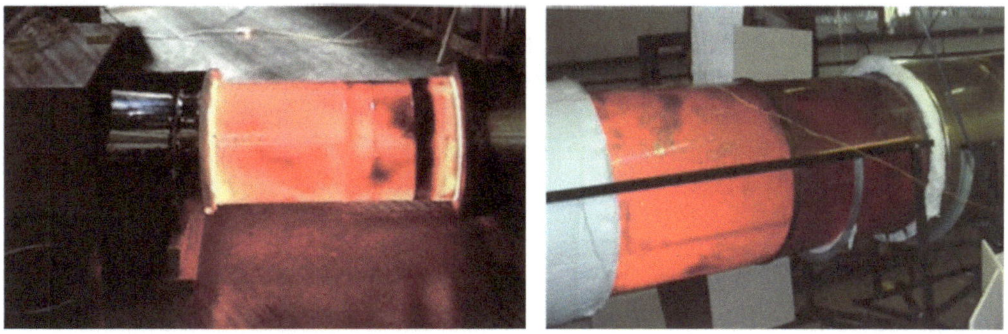

Figure 5.5 UL grease duct internal testing.

Courtesy of Selkirk Corp/Hart & Cooley (c/o Duravent Group)

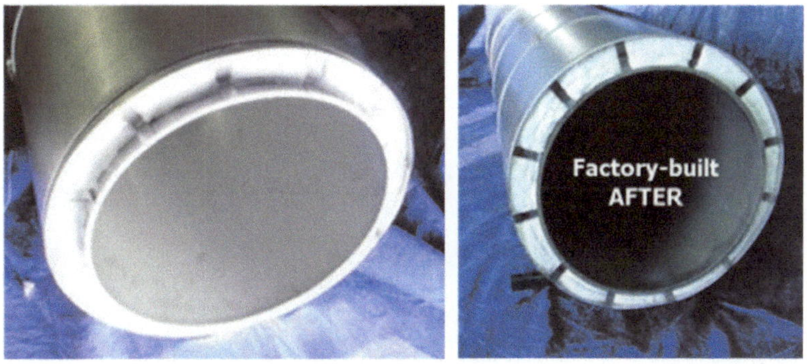

Figure 5.6 Listed round stainless steel duct before and after a UL severe internal test.

Courtesy of Selkirk Corp/Hart & Cooley (c/o Duravent Group)

Figure 5.7 Unlisted wrapped rectangular carbon steel duct before, during, and after a UL severe internal test.

Courtesy of Selkirk Corp/Hart & Cooley (c/o Duravent Group)

face roughness and can help save the building owner costs on cleaning maintenance, especially when automatic wash equipment is used (i.e., with rotating-head spray nozzles [see Figure 5.8]). Figure 5.9 shows approximately six months of grease deposition on a listed round stainless grease duct before and after cleaning. Rectangular ducts often require more manual cleaning labor and can pose a risk of unthorough cleaning if door locations/size/accessibility are limited (see Figure 5.10).

Installation Examples

Double-wall listed ducts have a metal outer wall to protect the insulation from incidental job-site damage. These products usually carry a factory warranty from the manufacturer. Typically, the outer wall is constructed of galvanized, aluminized, or stainless steel. In addition, the outer wall can be field or factory painted, if necessary. See Figure 5.11 for examples.

Field-applied flexible insulation grease wraps on unlisted carbon steel ducts have an aluminum foil face covering. See Figure 5.12 for examples of this type of duct.

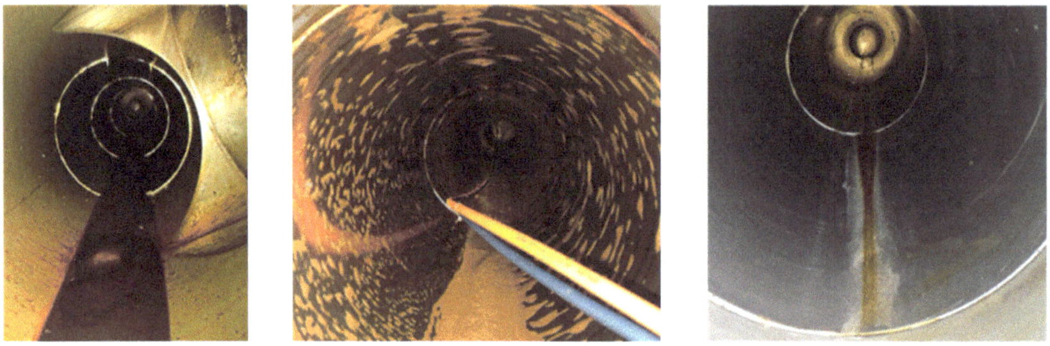

(a) (b) (c)

Figure 5.8 Rotating-head cleaning equipment in round listed duct: (a) typical grease deposition in round listed stainless grease duct, (b) mechanical detergent cleaning process in round listed stainless grease duct, and (c) post-cleaning results in round listed stainless grease duct.
Courtesy of DPW Services

(a) (b)

Figure 5.9 (a) Approximately six months of grease deposition on listed round stainless steel grease duct, and (b) the same duct post cleaning.
Courtesy of Enviromatic Corporation of America, Inc.

Figure 5.10 Limited access for manual scraping in unlisted rectangular duct.
Courtesy of DPW Services

Figure 5.11 Double-wall listed ducts have a metal outer wall to protect the insulation from incidental job-site damage.
Courtesy of Selkirk Corp/Hart & Cooley (c/o Duravent Group)

(a) (b)

Figure 5.12 (a) Field-welded duct and (b) field-welded and wrapped duct.
Courtesy of Selkirk Corp/Hart & Cooley (c/o Duravent Group)

6

Occupant Safety and Comfort Considerations

A successful kitchen requires the employees feel safe and comfortable in the fast-paced environment in which a kitchen can operates. A commercial kitchen is a potentially hazardous area with cooking equipment that may be supplied by propane or methane. Without a properly design commercial kitchen ventilation (CKV) system, the combustion byproducts of this equipment can have a major impact on the employees' health and could potentially affect patrons in the dining space. Additionally, excessive heat and moisture can lead to more than just discomfort; these factors can have adverse health effects on a person.

Employee illness can increase absenteeism, which has a negative effect on a kitchen's operation. It is estimated that employee absenteeism can cost a restaurant over 15% of the company's profit each year (Johnson-Tate 2018). Similarly, poor ventilation can impact turnover, as employees seek a safer, healthier, and more comfortable work environment. Additionally, both an unsafe and simply uncomfortable environment can keep patrons from returning.

This chapter brings awareness to common health hazards in kitchens and discusses ways to prevent these from occurring through proper kitchen ventilation design. It also focuses on comfort considerations for all occupants of the space.

Carbon Monoxide (CO)

CO Risk in Kitchens

Designers should be aware of safety when a kitchen uses gas appliances that, without proper design and safety measures, can release carbon monoxide (CO) into the space. Because CO is odorless, colorless, and undetectable to the human senses, kitchen staff may not be aware they are being exposed. The initial symptoms of low to moderate CO poisoning are similar to those of the flu and include headaches, fatigue, shortness of breath, nausea, weakness, and dizziness. High levels of CO can result in more severe symptoms including mental confusion, vomiting, loss of muscular coordination, loss of consciousness, and sometimes even death (U.S. CPSC 2023).

In 2018, the CDC published a report on an industrial kitchen facility in Wisconsin. The report notes that CO poisoning kills approximately 400 persons annually in the U.S. and states that in the specific incident described in this report, 45 employees in the Wisconsin kitchen were treated for it. The report concluded in the following way: "Gas-burning appliances in industrial kitchen facilities are not common occupational causes of CO-related morbidity and mortality. However, improperly maintained and ventilated appliances can be a source of CO exposure. Because symptoms of CO poisoning are nonspecific, CO poisoning might be underreported" (CDC 2018, 786).

Preventing Illness Due to CO Exposure

The CDC recommends the following: "Adequate ventilation in areas at risk for CO buildup, routine maintenance of gas-burning equipment, and detectors that alert to potentially unsafe levels are the best ways to prevent CO poisoning. OSHA does not specifically require CO detectors in industrial kitchen facilities; however, employers are required to evaluate all potential airborne contaminants that present a health hazard" (2018, 786).

Improper ventilation that causes CO to be released in a kitchen may be attributed to a lack of proper exhaust volume or inadequate makeup air, which may create back drafting of combustion gases from cooking equipment. It is strongly recommended that designers lay out appliance ventilation as recommended by the equipment supplier to properly exhaust combustion gases. A properly operated kitchen does not need high ventilation rate, but proper exhaust-to-supply ratios and configurations are key to safe ventilation practices.

In 2015, the State of New York passed a law requiring every restaurant in the state to install and maintain CO detectors (GTM 2015). Although not required in jurisdictions everywhere, it is highly recommended that CO detectors and alarming devices are installed in commercial kitchens where gas is used as a fuel source. For more information on Threshold Limit Values (TLVs®) and Biological Exposure Indices (BEIs®), refer to the guidelines published by the American Conference of Governmental Industrial Hygienists (ACGIH 2023).

Excessive Moisture

Mold

Molds reproduce through spores that are usually too small to be seen by the human eye. These small spores can be airborne and infiltrate indoor air in any building. These spores can land on a wide variety of substances and begin to propagate when moisture and oxygen are present by digesting whatever they are growing on to survive. In buildings with excess moisture, mold growth is common. Figure 6.1 shows an example of extensive mold contamination. If left undiscovered or untreated, human exposure to mold growth can have a major negative impact on health and possibly lead to highly dangerous conditions (EPA 2022).

Many types of molds exist, and all molds have the potential to be harmful to humans. Molds can produce allergens that can trigger allergic reactions and even asthma attacks. Some types of mold are toxic to the body (EPA 2022). Mold can even end up in a person's bloodstream, which can lead to symptoms such as irregular heartbeat, heart inflammation, internal bleeding, and blood clots (SAFE 2018).

Water is needed for continued mold growth; therefore, uncontrolled moisture can become a major problem in buildings. Some moisture problems in buildings have been linked to changes in building construction practices during the 1970s, 80s, and 90s, which resulted in buildings that are tightly sealed but may lack adequate ventilation, potentially leading to moisture buildup (EPA 2022). On the other hand, in buildings that are not tightly sealed, insufficient replacement air supply may cause negative building pressure and can lead to infiltration of humid air.

Moisture Control Guidance for Building Design, Construction and Maintenance (EPA 2013) identifies the following additional HVAC-related moisture problems. This list is repeated verbatim from page 57 of that book.

- **Inadequate dehumidification by the HVAC system during humid weather.** The resulting high indoor air dew point can lead to condensation, near-condensation and mold growth. Comfort problems also are common when the indoor dew point is high because the relative humidity is also high. Occupants often demand lower thermostat settings in an

Figure 6.1 Extensive mold contamination of ceiling and walls.
Courtesy of riopatuca/Shutterstock.com

attempt to be more comfortable. This is counterproductive. Lowering the thermostat overcools the building, which increases the risks of condensation, excess moisture absorption and subsequent mold growth. Cold temperatures also further reduce comfort and increase energy costs.

- **Leaking return and exhaust air duct connections, leaking indoor air handler compartments operating under suction, and leaking return air plenums.** These leaks create suction in building cavities, pulling humid outdoor air into the building where it can condense moisture onto cool surfaces and support mold growth.
- **Leaking supply air duct connections and leaking indoor air handler compartments operating under positive pressure.** When the weather is hot or humid, these cold air leaks chill surfaces behind walls and above ceilings, creating condensation and supporting mold growth. During cold weather these same leaks can force warm, humid indoor air into cold cavities where it can condense and support mold growth and corrode structural fasteners.
- **Oversized cooling systems.** When cooling systems, rather than dedicated dehumidification components, are expected to control humidity, the traditional inclination of designers is to increase the cooling tonnage to remove the combined sensible and latent loads. But overpowered cooling systems have exactly the opposite effect. A large cooling system removes the normal sensible load very quickly. Then, to avoid over-chilling the space, compressors are shut off or chilled water flow rate is reduced before the coils can condense enough moisture to control humidity. In nearly all cases, oversized cooling systems do not solve humidity control problems—instead, they cause them.
- **Ineffective drainage of condensate collected inside the HVAC system or condensation on the outside of the system's cold and uninsulated surfaces.** Undrained and uncollected condensate leads to water leaks, drips and subsequent moisture damage and mold risk.
- **Failure to exhaust indoor humidity sources such as showers, bathrooms, spas, pools and kitchens, especially in residential buildings and sports facilities.** Humid air can migrate from these sources to cold surfaces, leading to condensation or near-condensation and subsequent mold growth and structural deterioration.

Such moisture issues can lead to expensive health problems. The cost associated with asthma treatment due to mold is approximately $3.5 billion in the U.S., according to Lawrence Berkeley National Laboratory (2007). This figure does not include all the other negative health conditions related to damp buildings that have been reported, the addition of which is thought to be sizable.

Moisture damage to the building itself is also costly. Some of these related costs directly absorbed by owner include the following (EPA 2013):

- Repair and replacement cost of damaged goods such as furniture or supplies
- Repair and replacement costs due to corrosion of building structure, wiring, and/or damaged materials
- Increased cost of insurance
- Downtime following damage and throughout the mitigation or repair process
- Possible costs associated with litigation

Mitigating Moisture in Design

Potential health concerns are an important reason to prevent mold growth and to clean up and remediate any existing indoor mold growth. It is impossible to eliminate all mold and mold spores in the indoor environment; however, mold growth can be controlled indoors by controlling moisture indoors. The HVAC design goals to prevent excessive moisture are as follows:

- Control the humidity brought into the space through replacement air, and design for a low enough dew point to prevent condensation on cold surfaces. This condensation can drip onto materials that readily absorb moisture, such as ceiling tiles or drywall, and propagate mold (EPA 2013).
- Ensure that duct connections are sealed and there are no major sources of humid outdoor air that can readily cause infiltration to the building, such as through leaking return air ducts or plenums or a negative building pressure. Failure to seal supply duct connections can also lead to condensation in untempered areas in the building that are rarely inspected, such as above the ceiling (EPA 2013).
- Ensure proper equipment and insulation are specified to prevent overflowing drain pans due to condensate from cooling coils or condensation buildup on surfaces that may experience a large temperature differential (EPA 2013).
- Ensure that exhaust ducts from commercial kitchens include the required duct insulation in order to avoid internal condensation, because the dew point in the exhaust air is likely to be above the cool indoor air temperatures in the unheated cavities and attics that the ducts pass through on their way to the outdoors.
- Ensure that high sources of humidity gain to the space are properly exhausted (EPA 2013). Particular attention should be given to adequate exhaust from dishrooms. ASHRAE Research Project (RP) 1778 (Livchak and Swierczyna 2020) studied heat and moisture loads in commercial kitchen dishrooms and showed that latent loads can be mitigated with larger overhangs and higher airflow rates than codes currently require.

Excessive Heat

Effects of Excessive Heat

There is a common saying: "If you can't take the heat, then get out of the kitchen." Taken literally, kitchens are known for being hot, but that is not a badge of honor. Contributors to hot kitchens include untempered makeup air and high heat loads added to the space but not fully exhausted or overcome by inadequate cooling processes. Kitchen temperatures can rise to more than 110°F

(43°C), particularly in front of certain appliances such as open pizza ovens, grills, and anything with an open flame (see Figure 6.2). Even if personnel do not work continuously in front of particular appliances, there may not be a means for the employee to cool off in the rest of the kitchen work area.

Environments with high temperatures can lead to major health problems including, but not limited to, heat cramps, heat rashes, heat exhaustion, and heat stroke. The body is built to perspire by increasing blood circulation when exposed to high temperatures in an effort to cool the body down. While this is happening, less blood is circulated to important organs, including the brain, which may cause dizziness, blurred vision, nausea, and eventual collapse. A person experiencing heat exhaustion could suffer brain damage if not treated in a timely manner. Heat stroke is another condition related to heat exhaustion that has more serious consequences. It occurs when the body stops sweating and therefore cannot dissipate heat. This causes the body temperature to rise quickly, which can lead to serious brain damage and cause death (WCF 2023). Additionally, high exposure to heat often leads to dehydration, which can cause acute or chronic kidney failure (Singh et al. 2016).

It is also worth noting that personnel sweating in the kitchen can lead to food contamination and therefore health hazards to the patrons.

Preventing Excessive Heat

A commercial kitchen designer should take care to design with the following goals in mind:

- The ambient temperature in the kitchen should be 75°F (24°C)
  - Properly assess the heat load to the space and provide the necessary cooling tonnage to overcome this
  - Ensure the equipment selected to provide cooling is capable of cooling to the space design temperature at any outdoor air temperature up to the design temperature of the locale
  - Design for adequate control of the systems in place to interact with each other and with the space to ensure the design space temperature is being met throughout

Figure 6.2 Personnel working in front of hot cooking equipment.
Courtesy of iStock.com/bgton

- Ensure the vent hoods are fully capturing and containing heat from cooking (see chapters 3 and 4 of this book for more information)

Indoor Environmental Quality (IEQ)

Poor IEQ in Commercial Kitchens

In addition to problems created by CO, moisture, and heat when a space is not properly ventilated, the risk of other harmful particulate is high in a commercial kitchen system. Grease-producing cooking methods are shown to increase particulate matter (PM), volatile organic compound (VOC), and polycyclic aromatic hydrocarbon (PAH) emissions in commercial kitchens' indoor air. Studies have shown this can lead to kidney dysfunction, cancer, or other harmful side effects (Singh et al. 2016).

Other organic and reactive chemicals that contribute to indoor air contamination (such as tobacco smoke, vehicle exhaust, fugitive emissions, and more) may come from the outside the space either through infiltration into a negatively pressurized building or through the building's ventilation system and sources of outdoor air (OSHA 2023).

While health concerns are a primary driver for improving indoor environmental quality (IEQ), comfort requirements add another layer to the IEQ discussion. Thermal comfort is shown to affect workplace performance and productivity. One study in particular showed that a 10°F (5.6°C) rise in temperature led to a 30% loss in productivity. With cooking equipment and processes that produce high heat and moisture along with the exchange of a high volume of outdoor air, commercial kitchens are prone to space conditions that are less than ideal. Within a growing commercial kitchen industry with broad opportunity available, employee and patron retention is of rising importance to the survival of a particular concept, which makes comfort an important part of the IEQ equation (Simone et al. 2013).

ASHRAE RP-1469 (Stoops et al. 2013) evaluated over 100 commercial kitchens in the U.S. during the summer and winter in all climate zones. This study employed the commonly used predicted mean vote (PMV) index to evaluate indoor thermal comfort, which is recommended only for values between –2 and +2 on the seven-point PMV scale (ASHRAE 2020). Figure 6.3 shows that several data points of PMV for climate zones were outside the recommended range of evaluation, which clearly suggests a high percentage of dissatisfaction among kitchen employees. The authors suggest that "the large difference in the thermal environment between summer and winter in casual kitchen types was probably due to more frequent use of natural ventilation" (Simone et al. 2013, 1014) indicating a lack of mechanical heating and cooling in those environments. See Figure 6.4 for the operative temperature data.

Improving IEQ in Commercial Kitchens

Each of the following are ways to vent out harmful pollutants created in a commercial kitchen space and simultaneously prevent pollutants from entering the space through the outdoor air being brought into the building:

- Proper exhaust and outdoor air exchange
- Removal of pollutants through the outdoor air supply unit using media filters (Figure 6.5)
- Removal of pollutants through the outdoor air supply unit using other technologies such as ultraviolet (UV) treatment of air and bipolar ionization (specifiers should evaluate the research behind such technologies, as there are currently no ASHRAE standards for testing these technologies)
- Humidity levels maintained between 40% and 60% rh, as humidity levels outside of this range can increase transmission of viruses (ASHRAE 2022a)

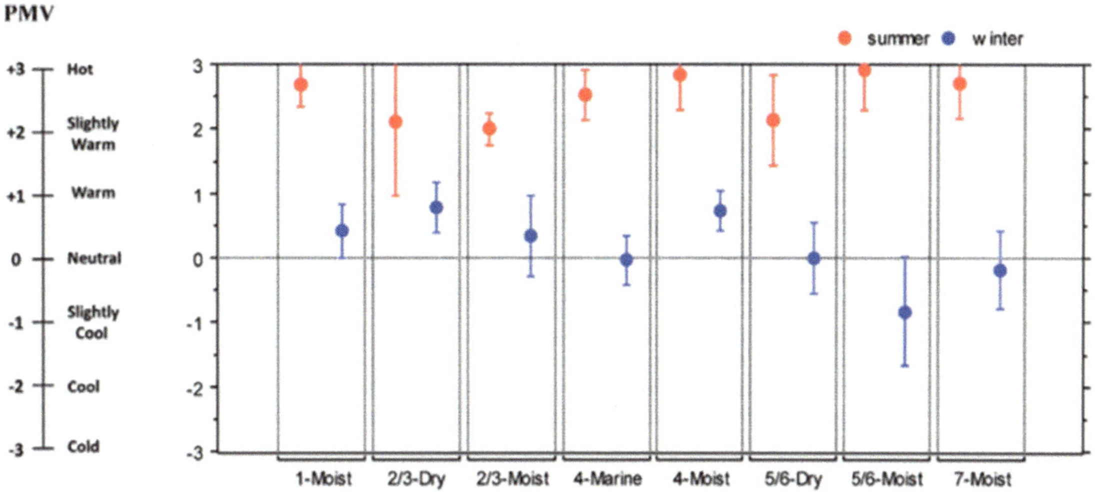

Figure 6.3 Average of PMV for climate zones with 95% confidence interval.
Courtesy of ASHRAE (Stoops et al. 2013)

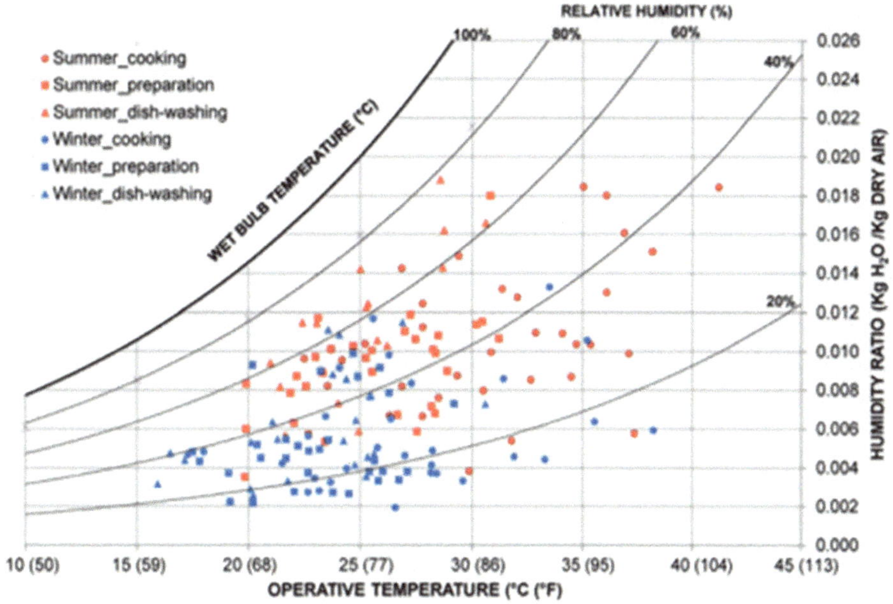

Figure 6.4 Operative temperature data from all kitchen zone short-term measurements in summer and winter.
Courtesy of ASHRAE (Stoops et al. 2013)

- Proper pressure control to prevent infiltration as well as issues in ventilation of necessary contaminants or excessive heat, moisture, or smoke

To improve thermal comfort, consider the following:

- Determine which outdoor design conditions should be used for the project location
- Properly calculate the energy load to the space and how much exhaust air, supply air, and tons of cooling are needed to keep the space comfortable

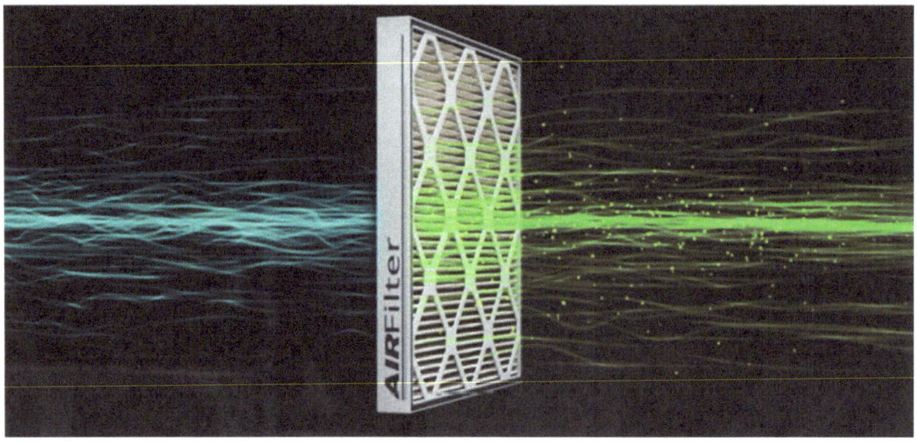

Figure 6.5 Depiction of harmful pollutants being filtered with the use of media filters.
Courtesy of Accurex, a Greenheck Group Company

- Carefully select equipment with enough heating and cooling capacity to overcome the load on a typical design day
- Carefully consider the selected equipment's capabilities during part-load conditions

See chapter 7 for more information on accounting for thermal comfort in the design process.

7

Comprehensive Design Approach

As discussed in chapter 2, the design of a commercial kitchen ventilation (CKV) system begins with the complete knowledge of how the facility is being used. Understanding of the Location of the kitchen, type of kitchen/restaurant being designed, type of food/menu being prepared, and hours of operation are key considerations. CKV design should also consider the appliances required to meet the needs of the kitchen space and the overall layout of the kitchen. Finally, it is important to consider how the kitchen layout and CKV considerations relate to the rest of the building's HVAC requirements. Figure 7.1 shows a design process map.

Understanding the CKV System's Overall Requirements and Constraints

Menu and Food Preparation

The kitchen's desired menu and cooking processes play a large role in the appliance selection and directly affect CKV design. See chapter 3 for a detailed discussion of the equipment used.

The design engineer should have clear expectations and understanding from the owner of the goal of the kitchen space. This critical information is necessary to select the required appliances and to determine the ventilation hood's airflow, the replacement air quantity, and the required HVAC equipment and devices necessary to condition the spaces. The space's occupancy limits, along with the hours of operation, should also be considered to guarantee accurate CKV design. Relating to hours of operation, it is important for the designer to consider how ancillary spaces will be handled, because other spaces may have separate HVAC systems and operating schedules.

There are four common hours of operation used in food service—breakfast/brunch, lunch, dinner, and late night. Refer to chapter 2 for definitions. To design a CKV system, the designer should know the hours of operation to determine the schedule for all HVAC systems in the facility. The designer also needs to fully understand the diversity of the food preparation equipment's operation so the heat gains can be accurately calculated. Finally, an understating of the impacts of all associated equipment on the HVAC system's design and operation must be considered. Note the following critical design criteria:

- Optimizing the hood's exhaust requirements, while maintaining capture and containment (C&C), minimizes the associated replacement air quantity, which in turn reduces the size of the required HVAC equipment as well as the associated operating costs.
- Similarly, minimizing the heat gains from food processing equipment can further reduce HVAC equipment sizes and operating costs. Methods for analyzing these costs are described in chapter 1.

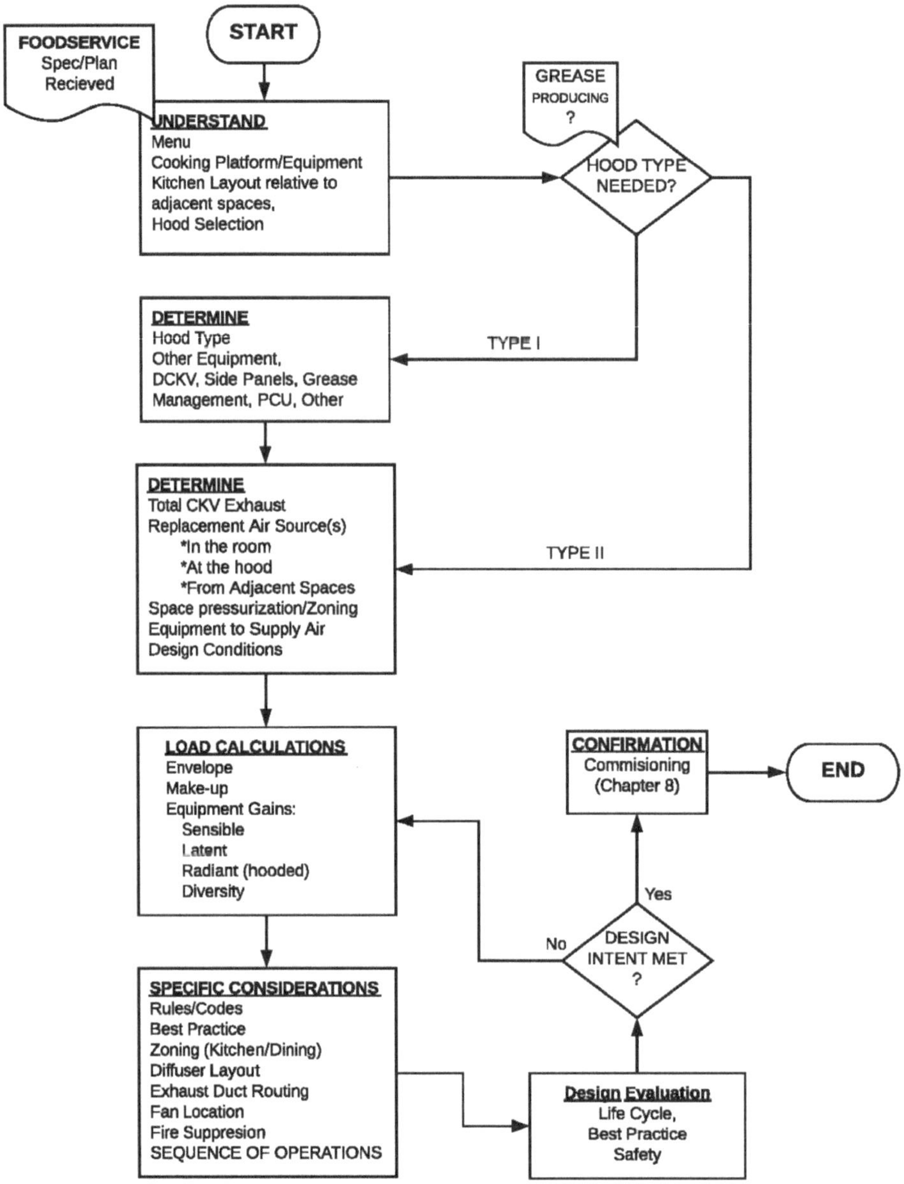

Figure 7.1 Design process flowchart.

Courtesy of Gaylord Industries

Location and Building Types

It is important to identify the building type, as this will influence the pressure conditions of the space, equipment locations, and duct routing. Chapter 2 of this guide defines the different building types of kitchens and restaurants. For the purposes of this chapter, they are listed here:

- Quick-service restaurants (QSRs)
- Fast-casual restaurants
- Full-service restaurants

- Ghost kitchens
- Tenant restaurants
- Other restaurants
- Food trucks

The physical layout and restaurant type can vary drastically relating to the design of the space. A QSR's dining and kitchen areas may only be separated by a serving counter. Many times, this type of restaurant is a single open area. In other restaurants, such as fast-casual dining restaurants, a wall may separate the dining and kitchen areas, creating two separate conditioned areas. For this example, the designer could keep the two conditioned areas separate and use demand ventilation controls in the dining space or use the maximum amount of transfer air by integrating the HVAC with the kitchen ventilation system.

Freestanding restaurants, in-line restaurants, and other restaurants where the restaurant occupies the entire building or the entire enclosed tenant space are often easier to design, as the engineer has full control of the building design. Tenant restaurants can be more difficult to design due to pressure relationships with common areas and the need for supplemental cooling and replacement air beyond what the landlord may provide. Tenant restaurants are typically designed with a negative pressure, as they use transfer air from the common area. However, if transfer air is not available, there will need to be 100% replacement air introduced to the kitchen to maintain pressurization.

It is important that the engineer also collaborate closely with the architectural designer when laying out the space. In many QSRs and fast-casual restaurants, an open concept where the cooking area is fully open to the dining area is typical. This type of design approach is desirable, because it allows air to easily transfer from the dining area to the kitchen area. While there may not be a detectable air pressure difference between the two areas, the quantity and velocity of the transfer air can be calculated during the design and quantified during air balancing and commissioning.

System Integration and Design

System integration and balancing bring the many ventilation components together to provide the most comfortable, efficient, and economical performance of each component and of the system as a whole. In CKV, the replacement air system(s) must integrate and balance with the exhaust system and/or facility HVAC system(s). Even optimal system designs require field testing and balancing once installed. It is equally important to verify compliance with design and to confirm that the design meets the operating facility's needs. Air balance is a critical step in any CKV Commissioning Process. It should be noted that a majority of hood performance issues are caused by improper balancing and placement of the CKV systems. Although there are exceptions, the guidelines below are the fundamental principles of integrating and balancing food service facility systems for comfort, control, and economical operation:

- Pressure conditions. Due to the large rates of exhaust air, makeup air, etc., there is a need to design for desired pressure conditions. The pressure conditions are as follows (see Table 7.1 for a list of recommended pressure relationships):
  - Positive air pressure—created when more outdoor air is supplied to the space than is removed from it
  - Negative air pressure—created when more air is exhausted from the space to the exterior than outdoor air is supplied to it
  - Neutral air pressure—created when the outdoor air supplied and the exhaust air are nearly equal

Table 7.1 Pressures for Various Space Relative to Other Spaces

Space/Room	Pressure Relative to Adjacent Spaces
Kitchen, cooking	Negative (–)
Kitchen, prep	Negative (–)
Storage	Positive (+)
Dish washing	Negative (–)
Offices, break rooms	Positive (+)
Dining rooms	Positive (+)

- The building should always be slightly positively pressurized compared to atmosphere to prevent infiltration of outdoor air. Infiltrated air contains contaminants and insects and adds to the heating and/or cooling load.
- Every kitchen should always be slightly negatively pressurized relative to adjacent rooms or areas immediately surrounding to help contain odors.
- System HVAC design should prevent air supplied to the kitchen from being returned and supplied to non-kitchen areas. Odor contamination is an obvious potential problem. In addition, in conditions such as seasonal transitions, when adjacent zones may be in different modes (e.g., economizer versus cooling or heating), comfort may be adversely affected. Ideally, the kitchen HVAC system should be separate from all other zones' HVAC systems. The following situations should be considered:
    - During seasonal transitions, the kitchen zone, due to its large internal heat gain from the cooking appliances, may require air conditioning or may be served by ventilation air only, while dining areas require heating. Even in kitchens that require cooling when the adjacent dining areas require heating, it is still important to maintain the pressure differential between these spaces and continue transfer of dining-area air into the kitchen.
    - To limit kitchen personnel discomfort, it is important to control the low-temperature makeup air set point and prevent drastic temperature variations between the kitchen space and the makeup air being introduced. When dedicated kitchen makeup air requires heating, thermostatic control of the makeup air heating source should ideally be based on kitchen space temperature rather than outdoor air temperature. makeup air heating should be interlocked with kitchen HVAC cooling to prevent simultaneous heating and cooling. HVAC thermostat locations in kitchens must consider the potential conflicting temperatures.
    - Ideally, in dining areas there should be no perceptible drafts and no temperature variations of no more than 1°F (0.6°C). These conditions can be achieved with even distribution and thorough circulation of air in each zone by selecting an adequate number of grilles, registers, or diffusers to preclude high air velocities in the occupied zone. If there are noticeable drafts or temperature differences, dining customers will be uncomfortable and facility personnel are generally less comfortable and less productive.
    - It is not recommended to use a return air plenum in a kitchen's HVAC design due to the possible accumulation of oil and grease on the surfaces and equipment located above the ceiling (e.g., lighting and wiring). Due to the operational nature of a kitchen, there will always be some quantity of grease and oil that is suspended in the return air;

however, while the areas below the ceiling can be easily cleaned, the above-ceiling areas cannot be cleaned and will accumulate to a substantial buildup after a few years of operation. For this reason, it is recommended to use a dedicated return air duct system.

Both design concepts and operating principles for proper integration and balance are involved in achieving desired results under varying conditions. The same principles are important in almost every aspect of food service ventilation.

In restaurants with multiple exhaust hoods, or hoods with demand control systems, exhaust airflow volume may vary throughout the day. Replacement air must be controlled to maintain proper building and kitchen differential pressures to ensure the kitchen remains negative to adjacent areas at all operating points while providing adequate replacement air for the operating hoods. The more variable the exhaust, or the more numerous and smaller the zones involved, the more complex the design, but the overall pressure relationship principles must be maintained to provide optimum comfort, efficiency, and economy. See chapter 4 for more information on design approaches.

Selecting Exhaust Hoods

A critical step when designing commercial kitchens is selecting which types of exhaust hoods to use. This step must be performed prior to the other steps necessary to design the HVAC system, because the hood will impact the amount of outdoor air needed, which in turn has a sizable impact on the HVAC load calculations and the associated equipment. This section describes the kinds of hoods available and the impact of hood location on hood exhaust airflows, how to determine the design exhaust airflow and evaluate the options available to provide reductions in both the design airflow and real-time energy savings, and other factors that need to be considered. There are two types of hoods available for CKV systems:

- Type I hoods are required when the cooking processes produce grease, heat, smoke, steam, and combustion byproducts (collectively referred to as *cooking effluent*).
- Type II hoods may be used when equipment produces only heat and steam.

It is recommended that the hood supplier(s) be consulted regarding any of the design criteria discussed in this section so that an optimal selection can be made. It is also strongly recommended to investigate and evaluate the performance of different hood suppliers' products for the same application, because variations between manufacturers can produce different results, including different C&C exhaust airflow quantities for covering the same cooking appliance(s).

It should be noted that when properly designed, exhaust hoods capture the sensible and latent convective loads from the cooking processes. However, the cooking processes can still add large radiant loads to the kitchen from their hot surfaces, and these radiant loads need to be accounted for when sizing the HVAC system for the kitchen. See the "Calculating Space Heat Gains and Losses" subsection of this chapter for more information on the process for determining HVAC loads. See chapter 3 for more information on the different types of exhaust hoods.

Styles of Exhaust Hoods

There are several styles of Type I exhaust hoods available for use in commercial kitchens, but the most common types are canopy hoods and backshelf hoods. The main differentiator between these two types is the installation height. Canopy hoods are typically installed so the lower edge of the hood is 78 to 84 in. (1.98 to 2.13 m) above the finished floor. This allows clearance for taller appliances, such as overfired broilers and tall combi ovens. Proximity hoods are typically installed closer to the appliances' surfaces and therefore require less exhaust air to capture the effluent from the appliances underneath the hood. Examples of appliances that commonly use proximity hoods include deep-fat fryers and flat-top griddles. With proximity hoods, care needs to be taken with some appliances to make sure that there is sufficient clearance when the appliances have parts or features that move manually or automatically—appliances that could have a clearance issue include fryers with automatic lift baskets or clamshell grills that have an upper platen that could hit the front of the hood if there is insufficient clearance. It should be noted that there are also hybrid hoods that are installed at a height between canopy and proximity hoods.

Type II hoods are most commonly used to capture heat and vapor but not grease or smoke. Type II hoods used to capture condensation are often designed to capture and drain the condensed water to floor drains or condensate gutters on the hood's perimeter where the water accumulates and is evaporated over time back into the hood's exhaust. Type II hoods are not covered under a product listing category nor do they carry a fire rating; therefore, they are more simplistic to design than Type I hoods. The Type II hood design process focuses primarily on hood overhangs and designing adequate ventilation rates. Note that the minimum airflow for a dishwasher shown in code may not be high enough for the dishwasher type being used in practice. Consult the hood manufacturer for design guidance. When conditions allow for the use of a Type II hood, it can be an attractive option in the design process due to the sometimes significant reduction in system cost over Type I hoods, which require grease filters, grease ducts, and grease-rated exhaust as well as a full fire suppression system.

Cooking Duty

The next item to be considered when determining a hood's exhaust airflow is the appliances' cooking duty. There are four classifications of duty level in *International Mechanical Code* (IMC; ICC 2021d): light duty, medium duty, heavy duty and extra-heavy duty. The higher the duty level, the greater quantity of air the hood must exhaust to adequately capture and contain the appliance's cooking effluent. ANSI/ASHRAE Standard 154 (2022d) contains recommended duty levels for several appliance categories. Be aware that, ultimately, the local authority having jurisdiction (AHJ) has the final determination in how the appliances might be classified with regard to cooking duty. Typical classifications of appliance duty levels are shown in Table 7.2.

Hood Optimization

Another consideration when selecting a hood is determining the degree to which the design exhaust airflows should be optimized. There are a number of ways to optimize the hood's exhaust airflows.

Appliance location. Optimization can occur by changing the appliance layout to relocate the heavier-duty-level appliances so they are in the middle of an appliance lineup and the lighter-duty appliances are located to the left- and right-hand sides of the exhaust hood. This generally allows the exhaust hood to operate at a lower exhaust airflow and offer a more reliable system for C&C of the cooking effluent.

Adding side panels. A second means of reducing the exhaust airflow is to add side panels to the ends of the exhaust hoods. These panels reduce the amount of room air being entrained into the exhaust air, leading to a smaller thermal plume from the cooking appliances, as shown in Figure 7.2. Some hood styles (i.e., backshelf hoods) may have side panels integrated into the hood design, whereas other hood styles (i.e., canopy hoods) may offer side panels as an option. It should be noted that different side panel designs provide different levels of exhaust airflow reduction.

Hood type and location. A third means of reducing the exhaust airflow is to pay attention to the hood location. An island-style hood (i.e., a hood located away from any walls) entrains air from

Table 7.2 Typical Appliance Duty-Level Classifications

Duty Level	Appliance
Light	Oven Range top
Medium	Fryer Griddle Smoker Electric broiler
Heavy	Gas broiler Plancha grill
Extra-heavy	Solid-fuel cooking appliance

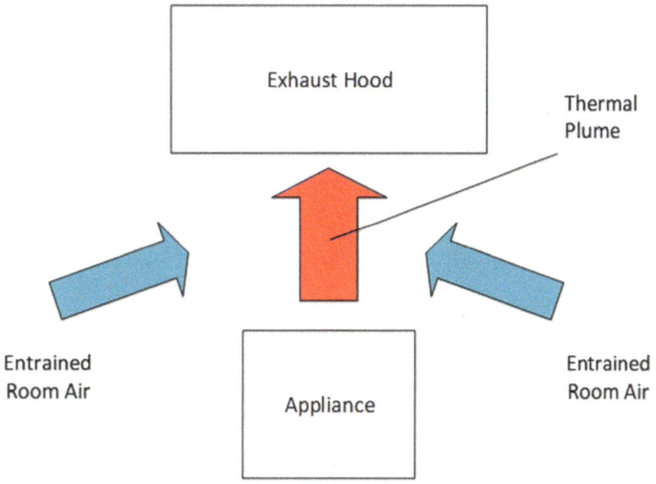

Figure 7.2 Appliance thermal plume and entrained air.

all open sides of the hood, which mixes with the plume from the cooking appliance, leading to higher required exhaust airflows (see Figure 7.2). A wall-mounted hood without side panels entrains air from three sides, whereas a wall-mounted hood with side panels or a proximity hood entrains air only from one side. Another option to reduce airflows would be to place the exhaust hood in a corner of the room to reduce the amount of room air being entrained.

Additionally, as the distance between the lower edge of the hood reservoir and the appliance cooking surface increases, the amount of air entrained from the surrounding space increases, resulting in larger exhaust airflows required to capture the cooking effluent. Stated simply, a hood installed closer to an appliance reduces the amount of required exhaust airflow and reduces the number of open sides. See Figure 7.2 for an example of how room air is an integral component to ensure C&C of a cooking appliance's thermal plume.

Hood Dimensions

When determining hood dimensions, the minimum left, right, and front overhangs (or underhangs) relative to the cooking equipment need to be taken into account. Most hood manufacturers have a listed minimum for overhang for each hood model, but a possible best practice is to exceed this minimum based on the appliance types or space conditions.

Calculating Exhaust Airflows

After the appliance lineup has been defined, the exhaust airflow needs to be calculated for the exhaust hood. Most kitchen ventilation hood manufacturers have in-house design tools to calculate the required exhaust airflows for a given cooking appliance lineup and associated hoods. These software packages use various calculation methods such as updraft velocity and thermal plume to determine the quantity of exhaust air required for the hoods. It should be noted that the calculated design airflow is generally higher than the hood manufacturer's minimum listed airflows (driven by the highest duty level for the appliances underneath the hood). The minimum listed airflows should not be used as the basis for design in the field.

Manifolding Hoods

When a commercial kitchen has more than one exhaust hood, a common design practice is to manifold branches of ductwork together to a common (or shared) duct and fan system. When combining such hood systems, the engineer needs to calculate the largest static pressure drop of all the branches (hood plus duct) and use that value when determining the overall static pressure drop the fan will see. One common mistake is to sum the static pressure drops from each branch together, which will result in an oversized fan ventilation system. Refer to chapter 4 of this guide for more information on static pressure in exhaust ventilation systems.

Demand-Controlled Kitchen Ventilation (DCKV) Systems

Once the design of the hood's exhaust airflows has been optimized for the cooking operations, the next step is determining whether to use a demand-controlled kitchen ventilation (DCKV) system to reduce energy consumption.

The basic premise of DCKV systems is that when the cooking appliances are in an idle state, the hood exhaust airflows are at a minimum. When any appliance underneath an exhaust hood is cooking, the hood exhaust should increase to the design airflow for that hood to properly capture the heat, smoke, grease, and other effluent from the cooking process.

The mechanisms used to determine the state of the appliances underneath the hood vary in complexity. These range from a single temperature sensor in the exhaust collar to more complex systems that use multiple temperature sensors in the reservoir of the exhaust hood to systems that additionally use optical detection devices.

It should be noted that as the exhaust airflows regulate up and down, the replacement air to the kitchen also needs to regulate to maintain proper kitchen pressurization. The energy savings from DCKV systems are consequently a combination of the exhaust fan energy savings and the replacement air cooling and heating energy savings, depending on what types of systems are used to condition the replacement air coming into the kitchen space.

One of the allowable and more achievable options to comply with the ANSI/ASHRAE/IES Standard 90.1 (ASHRAE 2022b) requirement for kitchens exhausting more than 5000 cfm (2360 L/s) and the IgCC/189.1 (ICC 2021a) requirements for kitchens exhausting more than 2000 cfm (950 L/s) is that DCKV must be capable of at least 50% reduction in exhaust. Caution should be used when selecting a demand-controlled system for some types of equipment. If all the equipment under the hood is unlikely to change in total load throughout the day (i.e., a single donut machine that only runs during peak breakfast hours), a DCKV control system is not warranted. For equipment that rapidly changes effluent conditions under the hood (i.e., a wok), a DCKV system (both the sensors and the fans) that can just as rapidly respond with increased exhaust should be selected.

Commercial Kitchen Air Design

Maintaining Air Balance in the Kitchen

Determining air balance for a kitchen should begin with the volumetric airflow rate (cfm or L/s) required to properly exhaust all effluent that is created from the appliances under the hood at maximum cooking capacity. There are different methodologies for these calculations. Unlisted hoods should reference IMC Chapter 507 (ICC 2021d) for determining airflow. However, it is recommended that designers always use listed hoods for Type I applications and that they enlist the hood manufacturer's expertise to determine the proper exhaust volume necessary for the appliance lineup.

Next, assess all other sources of exhaust in the space. This may include general exhaust for a kitchen when it is occupied that is not directly related to cooking, such as toilet exhaust, storage area exhaust, etc. These values combined provide the total exhaust needed. Knowing that kitchens are negatively pressurized up to –0.02 in. w.g. (4.98 Pa) relative to adjacent spaces for odor control (though still positive relative to the exterior), transfer air from the dining space (or other spaces) can be relied upon for a portion of the replacement air. For an example of this air movement, see Figure 7.3. The volume of transfer air is determined primarily by two considerations:

- How much conditioned outdoor air is being supplied to the adjacent space to meet that space's occupancy requirements per ANSI/ASHRAE Standard 62.1 (2022c). Note that ASHRAE Standard 62.1 classifies kitchens as Class II spaces and as such allows the transfer of air from Class I spaces, as per Table 6-1.
- The ability to transfer air from the adjacent space (through a doorway, pass-through window, or open concept) while limiting any openings for the transfer air's velocity to be less than 50 fpm (0.25 m/s) as per typically recommended room-to-room HVAC transfer air design practices.

Once the maximum allowable transfer air into the kitchen has been determined, there are two further options for replacement air:

- Outdoor air brought into the space through HVAC units that are not tied to the kitchen control sequence of operations
- Outdoor air brought into the space that is directly tied to the kitchen control sequence of operations

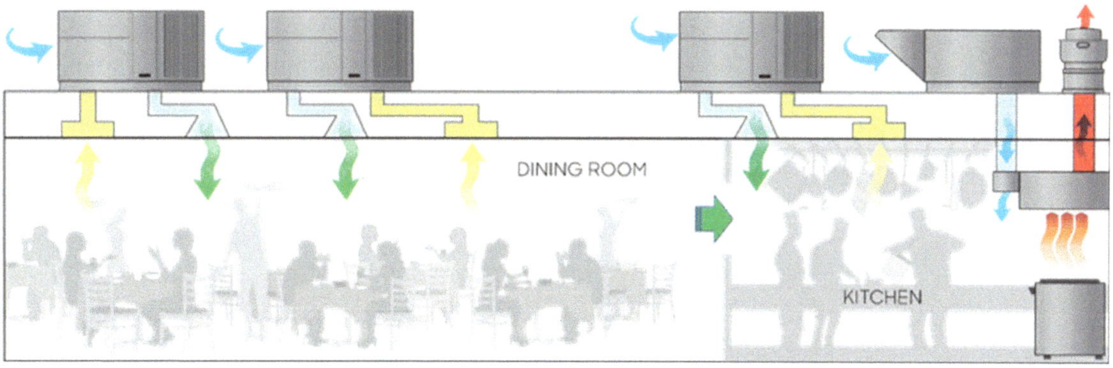

Figure 7.3 Example restaurant air movement.
Courtesy of Accurex, a Greenheck Group Company

Kitchen control sequence of operations refers to an HVAC unit or system that supplies an air volume that is directly proportional to what is being exhausted through the hood at any given time. If there is no replacement air directly tied to the kitchen controls in such a way, it is necessary to ensure that replacement air is properly supplied by other HVAC units that can provide replacement air while the kitchen exhaust fans run and ensure the design pressure is maintained at any given time. An example of this situation is when the kitchen is operating on a different schedule than the HVAC units that serve the spaces adjacent to the kitchen and which provide replacement air to the kitchen as part of the overall design to maintain the proper kitchen pressurization (i.e., slightly negative to the adjacent spaces).

It is also important to note that the level of conditioned air sometimes determines how it is introduced to the space. Typical makeup air for a CKV system is partially conditioned. Heating and cooling is recommended, but most often this air is not dehumidified, so it is brought in through plenums at or near the hood. The idea here is that this marginally tempered air should be the first to leave the kitchen space through the hood rather than the fully conditioned air in the kitchen inefficiently leaving through the means of exhaust. However, studies show that only about 50% is directly recaptured, and the other 50% becomes a load to the space (refer to chapter 34 of *ASHRAE Handbook—HVAC Applications* [2019] for more information). For this reason, dehumidifying this makeup air is a recommended design practice, because unlike hot or cold air, humidity diffuses throughout a space regardless of how or where it is introduced; therefore, its impact on space comfort cannot be controlled by air movement. For fully conditioned air (tempered and dehumidified), the intent is for this air to be distributed throughout the kitchen in such a way that the entire space is properly conditioned. Additionally, air velocities at the edge of the hood resulting from makeup air should not exceed 75 fpm (0.381 m/s) for any air distribution method selected.

Supply air distribution methods are discussed in detail in chapters 3 and 4, but to reiterate, it is important that directional diffusers and returns are not located within 15 ft (4.6 m) of the hood.

Equipment to Supply Air to the Kitchen

Most commonly, CKV systems use three types of equipment to supply air to the kitchen:

- Makeup air units (MAUs)
- Dedicated outdoor air systems (DOASs)
- Rooftop units (RTUs)

MAUs. MAUs are dedicated to providing makeup air to the exhaust hood to make up for much of what is being exhausted through the hood using 100% outdoor air. Usually, the total supply air volume through an MAU is less than the total exhaust air volume (this quantity is dependent on the other sources of supply and/or transfer air). Cold air sinks and hot air rises, which can contribute to C&C issues at the hood; therefore, it is recommended that this air be tempered to match the design temperature of the space, though it is rarely dehumidified. Since this particular type of replacement air is not dehumidified, it is most often brought in very close to the hood in such a way that it does not contribute to velocities at the hood edge exceeding 75 fpm (0.381 m/s).

MAUs are most often tied directly to the kitchen controls, so they proportionally track with what the exhaust fan is exhausting. When DCKV systems are used, it is particularly important that the MAUs have the capability to automatically turn down airflow to match the turn-down settings of the kitchen controls. For example, if the turn down is 50% (as is required if DCKV is the energy-saving method used per the *International Energy Conservation Code* [IECC; ICC 2021c] for kitchens with exhaust greater than 5000 cfm or 2360 L/s), the MAU should be capable of turning down to 50% of its design airflow during times of low cooking volume. This is more than a variable-frequency drive (VFD) or electrically commutated (EC) motor setting. For example, cooling coils and direct-fired gas burners require a certain velocity of air moving across them to ensure these tempering types work as designed.

DOASs. DOASs are typically used in a commercial kitchens when there is a need to fully condition a high volume of outdoor air. To accomplish full conditioning, the unit must have sufficient heating and cooling capability. To dehumidify, these units cool air well below the saturation point (typically between 45°F and 55°F [7.2°C and 12.8°C), thus condensing within the unit to strip moisture out of the air. Then, depending on the CKV system's design requirements, that air is reheated to the desired space conditions. An added benefit of DOASs over typical MAUs is the recirculation capability, which can lead to a cost savings on tempering air, help keep the space balanced, and allow for more flexibility in airflow. An example of added flexibility with recirculation capability is that many units have a lower limit on the design airflow in order to properly temper the air. There may be instances in which the kitchen exhaust fan and hood are not in use, but a small amount of outdoor air is still needed from the DOAS. With recirculation capability, the exhaust fan can still run at the lower required limit, but the desired amount of outdoor air can be met with reciprocal outdoor air and recirculation dampers within the unit.

It is recommended that the DOAS be tied directly to the kitchen control panel for proper operation while the hood and exhaust fan are in use, but there are many control methods available to keep the space balanced with a DOAS. Because the DOAS fully tempers the air, it is recommended that this air be evenly distributed throughout the space using appropriate distribution methods that will not lead to high velocities or turbulence in the kitchen.

RTUs. RTUs use a combination of outdoor air and recirculation and provide full tempering capability at design airflows. They typically do not supply more than 25% of their total design airflow with outdoor air. Because kitchens require a high volume of outdoor air, RTUs in kitchens are most often supplemented with an MAU or DOAS to replace much of what has been exhausted, though this is dependent on the exhaust volume and the amount transfer air able to be used.

Figure 7.4 shows a kitchen laid out with an MAU for directly making up what is exhausted from the kitchen hoods plus an RTU for full tempering throughout the kitchen and some additional outdoor air for replacement.

Figure 7.5 shows an alternative method of supplying air to the kitchen where a DOAS functions to both supply the bulk of the outdoor air needed for replacement and also fully tempering the space.

Design Temperature

ASHRAE Handbook—Fundamentals (2021b) publishes climatic design conditions for various weather stations (primarily corresponding to airport locations) throughout the world. Values for each station include annual heating, humidification, cooling, dehumidification, and enthalpy design conditions as well as extreme annual design conditions. Most equipment selections for replacement air methods require a design dry-bulb condition for both summer and winter and a wet-bulb condition for summer. Because ASHRAE offers climate data that correspond to different cumulative frequencies of occurrence, it is important to note that the design conditions used will affect the leaving air temperature of a unit that is selected for a given design condition. For example, using the ASHRAE-published cooling 0.4% annual cumulative frequency of occurrence will yield higher design temperatures—and correspondingly larger cooling equipment sizes—than using either the cooling 1% annual cumulative frequency of occurrence or the cooling 2% annual cumulative frequency of occurrence.

It is also important to consider part-load conditions, not just the leaving air conditions from a unit at the design temperature. All units and tempering types have a particular range within which the unit can properly apply tempering for airflow, humidity, and outdoor air temperature. The space design conditions should be able to be met at any outdoor air temperature or airflow that the unit may experience at any given time throughout the year. Verify with the unit manufacturer what

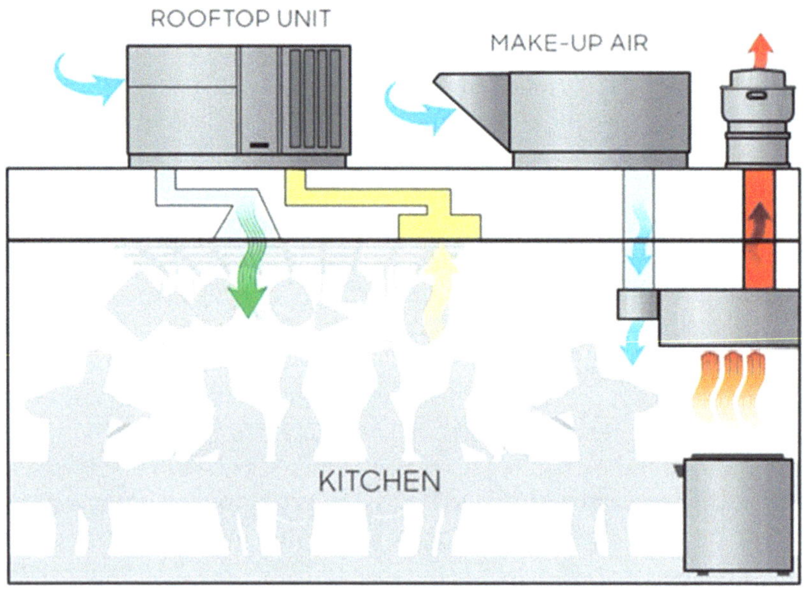

Figure 7.4 A kitchen where an RTU is used primarily for space tempering and an MAU is used primarily for replacement air for what is being exhausted through the hood.

Courtesy of Accurex, a Greenheck Group Company

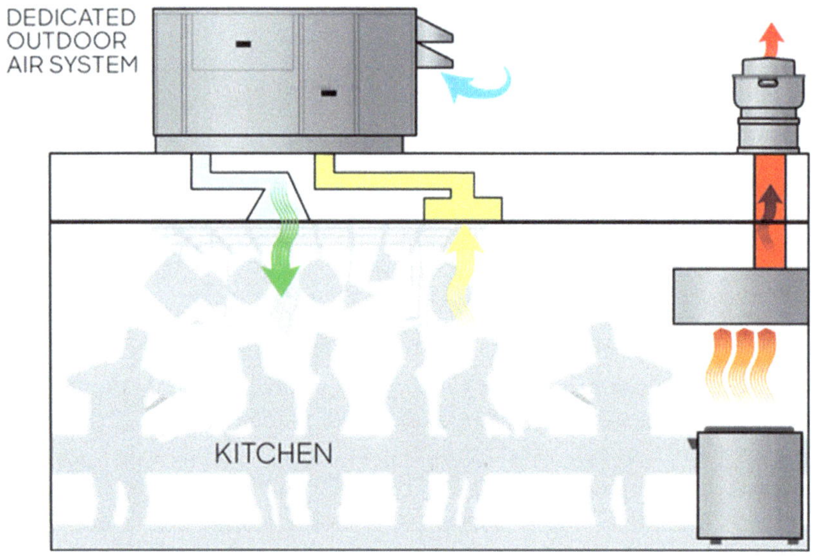

Figure 7.5 A kitchen where a DOAS is the only direct-supply source in the kitchen for full tempering and balance.

Courtesy of Accurex, a Greenheck Group Company

the tempering capabilities are for varying conditions. Note that extreme temperatures and climates can derate the unit's performance. Ensure this has been accounted for with the leaving air conditions defined for a supply unit.

Additionally, consider the climate zones and the length of time throughout the year that a supply unit's heating, cooling, and dehumidification processes will likely be operated. (Figure 7.6 shows the different zones for the U.S.).This may help in determining the cost-to-benefit ratio of a

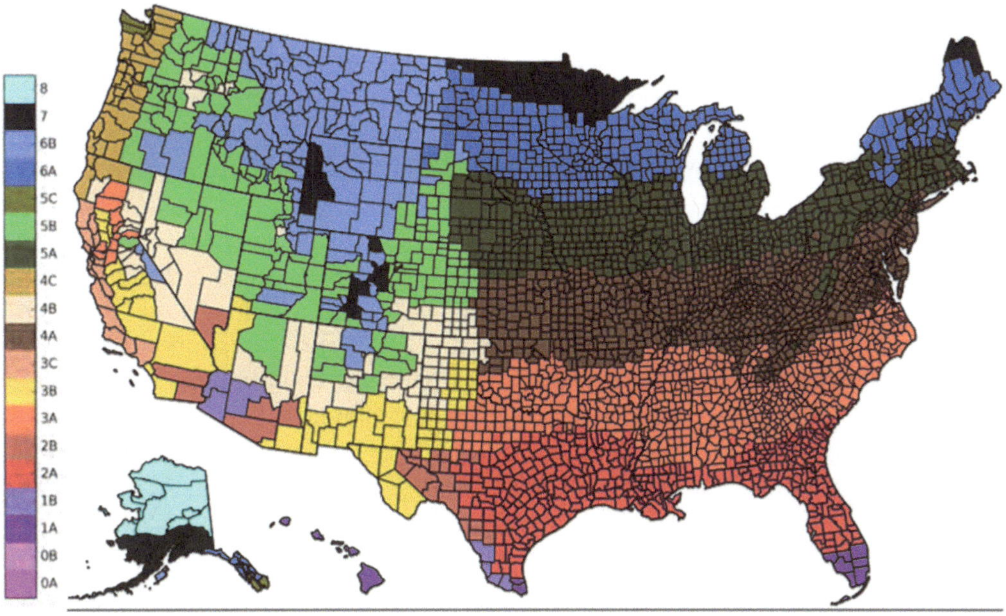

Figure 7.6 Climate zones for U.S. counties.
Courtesy of ASHRAE (2021a)

higher first-cost unit that provides full conditioning versus a unit that may not provide full conditioning during particular times of the year. For example, some units may only use their cooling capability for an estimated three days of the year based on that location's design conditions and the unit's inability to begin cooling until the outdoor air temperature reaches a particular set point. In this situation, it would not make sense to include cooling for that unit, or it may be more beneficial to select a different unit that can begin cooling at a lower outdoor air set point. Given that, the following should be considered when designing the system:

- For climate design conditions, using the 2% frequency of occurrence for cooling and 99% frequency of occurrence for heating is usually adequate. Kitchens are not mission-critical environments, so the extra up-front costs of HVAC systems to satisfy 0.4% cooling and 99.6% heating are difficult to justify.
- Given the large proportion of makeup air that a CKV system may require, when performing a cooling load calculation for a CKV system, it may be best to consider using the evaporative conditions for wet bulb rather than the cooling conditions, as use of evaporative conditions can sometimes require significantly more cooling to satisfy the design intent.

Load Calculations for Commercial Kitchens

This section assumes the designer has experience performing HVAC load calculations. Therefore, it focuses on the unique considerations that must be incorporated when performing these calculations for CKV systems.

Zoning

A commercial kitchen HVAC design typically needs to consider two separate zones: 1) a dining area, and 2) the kitchen areas, which consist of the cooking area(s), food preparations area(s), dish-washing area(s), and possibly the dry storage areas (e.g., nonrefrigerated storage areas). While each of these two zones may have similar hours of operation, their cooling and heating load profiles are likely to be very different.

Dining zone. The dining area, is typically served by an HVAC unit or system that is separate from the kitchen area. The unit typically handles the room loads including sensible, latent, and radiant. The loads from occupants and from the food being served should also be considered. Additionally, this would include the sensible and latent loads associated with the outdoor air intake necessary to meet ASHRAE Standard 62.1 (2022c) as well as to offset any exhaust from adjacent toilet rooms whose replacement air originates from this zone. It is recommended that if the dining zone is located within a larger building (such as a cafeteria in an office building), this zone should be served by its own HVAC equipment so that the dining area return air and the associated food odors from the dining zone will not mix with the return air from other non-dining-area zones.

The approach for performing HVAC load calculations for this zone is same as would be used for any typical occupied space and includes taking into account any transfer air from other adjacent spaces. See chapter 18, "Nonresidential Cooling and Heating Load Calculations," in *ASHRAE Handbook—Fundamentals* (2021b) for more information.

Kitchen zone. The kitchen zone operates very differently from the dining area, which is why it needs to have its own HVAC system. It will typically have a large sensible load (comparable to a data center), large latent loads (from dish-washing operations and food steamers), radiant loads (from the appliances located under exhaust hoods, which typically operate at several hundred degrees), and a large amount of outdoor air that must be conditioned. The HVAC system serving this zone also provides the required comfort conditioning for kitchen staff and food servers as listed in the "Excessive Heat" section of chapter 6. Engineers who are used to calculating HVAC loads for more traditional spaces may not realize how different the heat gains in a kitchen environment can be. Additionally, the operating schedule for a kitchen will likely be different than the restaurant's other zones or, if applicable, the attached building. Additionally, due to their large internal heat gains, there may be seasonal conditions when kitchens may require cooling while the adjacent zone requires heating.

Subzoning within the kitchen zone. When evaluating a restaurant's layout, the designer should identify area uses and occupancies to determine equipment zoning. Areas to consider include the following:

- Spaces with cooking and processing equipment
- Spaces with dish-washing equipment and other sources of excess moisture
- Spaces with heat-rejecting equipment such as refrigerators and freezers/coolers that do not reject heat to the building exterior
- Spaces with different activity levels (e.g., banquet rooms, meeting rooms, dining rooms, etc.)
- Staff break areas and offices
- Spaces with larger areas of glazing and doors to the exterior
- Interior spaces
- Storage rooms for food, paper, and packaging

Design plans for a facility with a CKV system should include a schedule or diagram that is created by the design engineer and that indicates the design outdoor air balance and overall air movement both within and relative to the spaces served by the CKV system. The design outdoor air balance should indicate all exhaust and replacement air for the facility plus the net exfiltration, if applicable. The total replacement airflow rate must equal the total quantity of air being exhausted by all exhaust fans plus a surplus amount to provide positive pressurization for the applicable spaces. Figure 7.7 shows an example of the airflows in a typical freestanding restaurant.

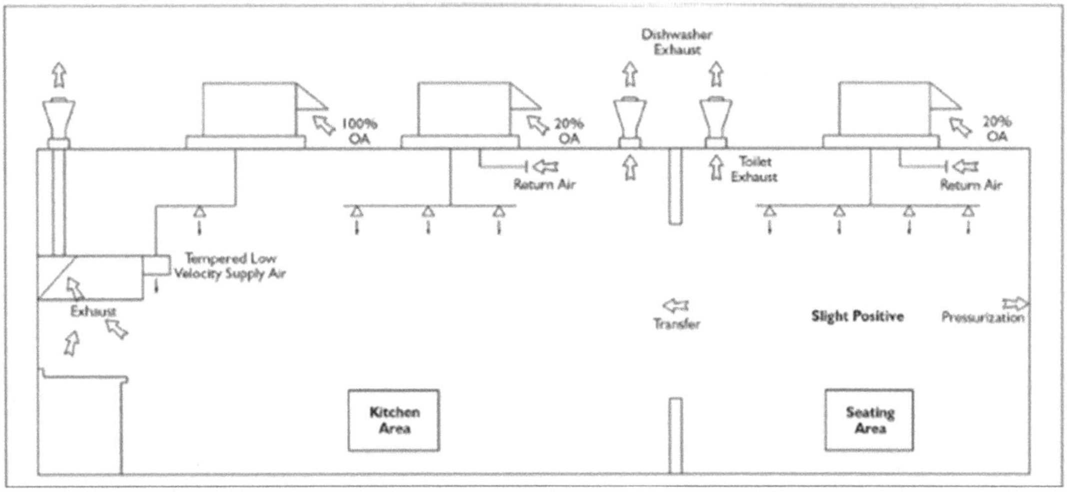

Figure 7.7 Typical CKV system airflows.
Courtesy of ASHRAE (Clark 2006)

Calculating Space Heat Gains and Losses

Given the large amount of outdoor air required and heat produced by cooking appliances, CKV systems have some of the highest cooling loads per floor area for any application, with the possible exception of data centers. In addition to the typical factors that are considered when performing CKV HVAC load calculations (envelope, lights, people), the considerable outdoor air quantity (which can be 50% or greater of the total load) and the heat gains from both hooded and unhooded equipment (which are typically substantially larger than the other internal heat gains) must be included.

Given that conditioning the required outdoor air is a large amount of the CKV HVAC load, it is recommended that the hood (and, if possible, the appliances under the hood) be selected to minimize the required exhaust. A slight change in hood design or configuration can have a large impact on the CKV system sizing. It is for this reason that evaluating hood designs and exhaust air reduction options should be a part of any proper CKV design process. This evaluation also applies when deciding whether or not to include DCKV.

Additionally, giving careful consideration to calculating the heat produced by the cooking appliances, including factoring their usage and diversity factor, is critical to performing an accurate CKV HVAC cooling load calculation. It is important to understand that not all cooking appliances will be operating simultaneously, and they will be operated as needed rather than continuously. It is recommended that the engineer consult with the end user or the kitchen facility designer to better understand the cooking appliances' intended hours of operation. That information can then be used to accurately calculate heat gains. Note that some kitchens' peak internal heat gains may occur during their breakfast or lunch hours. See Chapter 34 of *ASHRAE Handbook—HVAC Applications* (2019) and Chapter 18, Tables 5A through 5E, of *ASHRAE Handbook—Fundamentals* (2021b) for more information on different kitchen appliances' heat gains.

Note that cooking appliances that are located under an exhaust hood will not produce any direct-to-space sensible or latent heat gains, because that heat will be exhausted into the hood. However, they will produce a radiant heat gain to the space.

If the appliance usage diversity factor cannot be provided to the CKV HVAC designer, Table 7.3 can be used as a reference. It is recommended to multiply the heat gain from cooking appliances by the reduction factor, called the *simultaneous coefficient*, which is the quotient of the number of

Table 7.3 Recommended Values for Simultaneous Coefficient

Building Type	Kitchen Type Simultaneous Coefficient (KSIM) Range
Hotel	0.6–0.8
Hospital	0.5–0.7
Cafeteria	0.5–0.7
School	0.6–0.8
Restaurant	0.6–0.8
Industrial	0.6–0.8

Source: Tabunshchikov et al. (2020)

appliances in use by the total number of appliances. Table 7.3 lists coefficients for different building types.

Equation 7.1 calculates the sensible radiant heat gain to the space from appliances underneath a hood:

$$Q_r = K_{sim} \cdot P \cdot F_u \cdot F_r \tag{7.1}$$

where

Q_r = sensible radiant heat gain to space, Btu/h or W

K_{sim} = simultaneous coefficient for restaurant, from Table 7.3

P = appliance nameplate input, Btu/h or W

F_u = appliance usage factor (refer to the "Nonresidential Cooling and Heating Load Calculations" chapter in *ASHRAE Handbook—Fundamentals* [2021b])

F_r = appliance radiation factor (refer to the "Nonresidential Cooling and Heating Load Calculations" chapter in *ASHRAE Handbook—Fundamentals* [2021b])

Hooded appliance: sensible radiant heat gain example. For this example, assume there is a hooded underfired charbroiler with a nameplate input of 120,000 Btu/h (35,168 W) and that the kitchen is located in a hotel. For a hotel, the simultaneous coefficient ranges from 0.6 to 0.8, so use the average value of 0.7. From Table 5C in the "Nonresidential Cooling and Heat Load Calculations" chapter of *ASHRAE Handbook—Fundamentals* (2021b), the usage factor for an under-fired broiler is 0.84 and the radiant factor is 0.35. Using these values in Equation 7.1 results in the following equation:

$$Q_r = 0.7 \cdot 120{,}000 \cdot 0.84 \cdot 0.35 = 2496 \text{ Btu/h } (731 \text{ W})$$

To calculate the convective heat gain to the space for unhooded equipment, Equation 7.2 can be applied.

$$Q_c = K_{sim} \cdot P \cdot F_u \tag{7.2}$$

For this example, assume there is an unhooded baking oven with a nameplate rating of 25,000 Btu/h (7327 W), and the kitchen is located in a cafeteria. For a cafeteria, the simultaneous coefficient ranges from 0.5 to 0.7, so the average 0.6 is assumed for this example. The usage and radiant factors for this example are taken from Table 5A in the "Nonresidential Cooling and Heating Load" chapter of *ASHRAE Handbook—Fundamentals* (2021b). The convective heat gain to space is then calculated from Equation 7.2 as follows:

$$Q_c = 0.6 \cdot 25{,}000 \cdot 0.79 = 11{,}850 \text{ Btu/h (3473 W)}$$

The radiant heat gain to the space is calculated using Equation 7.1, or the result of Equation 7.2 can simply be calculated by the radiation factor as shown below:

$$Q_r = 11{,}850 \text{ Btu/h} \cdot 0.18 = 2133 \text{ Btu/h (625 W)}$$

The total heat gain to the space for the unhooded oven is the sum of the convective and radiant heat loads as follows:

$$Q_{total} = Q_c + Q_r = 11{,}850 \text{ Btu/h} + 2133 \text{ Btu/h} = 13{,}983 \text{ Btu/h (4098 W)}$$

Specific Design Considerations

Codes and Standards

One of the most important beginning steps in a CKV design project is to fully understand what municipality codes and enforcements (per the AHJ) the project will have to abide with to obtain a certificate of occupancy. Refer to appendix A of this guide for more information on the relevant codes and testing standards, including the following:

- ICC codes (*International Mechanical Code* and IgCC/189.1, *International Green Construction Code®*)
- NFPA standards (including NFPA 96, from which *Uniform Mechanical Code* often derives its text)
- ASHRAE handbooks and standards (including ANSI/ASHRAE Standard 154, *Ventilation for Commercial Cooking Operations*)
- UL standards
- ASTM standards
- *Uniform Mechanical Code*

Air Device Selection and Layout

Any supply-air device located within the kitchen or within proximity to a hood should be selected to minimize drafts and impact to hood C&C. Under no circumstances should directional or high-face-velocity devices be used. This includes, but is not limited to, supply grilles, perforated diffusers with directional vanes, linear slot diffusers, plaque diffusers, and cone diffusers. Refer to the "Impact on Hood Performance" subsection of the "Replacement Air" section of chapter 4 in this book for more information.

Type I Kitchen Hood Grease Exhaust Duct

A grease duct is either constructed by stainless steel listed factory-built designs (UL 1978 [2010d]) or unlisted field-welded methods built to minimum code construction requirements.

Grease duct routings should be as short as possible to the duct termination. This creates less static pressure and potential for less grease accumulation. It also creates long-term operational savings for the building owner with regards to reduced time for maintenance and system cleaning. Common industry design grease duct velocities are from 1500 fpm to 1800 fpm (7.62 m/s to 9.14 m/s) at design airflows. As with any duct-design protocol, consideration should be made for duct noise or excessive static pressure if the industry-design velocity ranges are significantly exceeded. Minimum velocities prescribed by the codes are 500 fpm (2.54 m/s). In buildings where it can be difficult to minimize the duct route up to the roof, an option is to discharge through an external wall with a grease mitigation system such as a pollution control unit (PCU).

The grease duct system should not be interconnected with any other building ventilation or exhaust system. If multiple hoods combine into a single common exhaust duct run, each junction fitting (i.e., a tee) and downstream duct should increase in size to maintain the preferred design duct velocity ranges.

Chapter 5 of this guide goes into further detail regarding grease exhaust duct-design options, hourly-rated enclosures, and other requirements and suggestions.

Exhaust Fans

Exhaust fans used to exhaust grease-laden vapors should be listed and labeled in accordance with UL 705, Supplement SC (2017b; formerly UL 762). Exhaust fans can be roof mounted, wall mounted, or in line. Fans should be located and installed in accordance with manufacturer recommendations as well as with NFPA 96 (2021b). More information on exhaust fans can be found in chapter 4.

Fire Suppression

This section highlights the primary considerations for designers when it comes to fire suppression systems for kitchen hoods and other ancillary products where it is required. For more in-depth detail on fire systems, please see chapter 3.

Fire suppression systems are required by IMC (ICC 2021d) and NFPA 96 (2021b) for any Type I hood system, meaning any hood that covers grease-producing equipment. These required fire suppression systems are typically wet chemical or water-based systems that are piped in the hoods to deliver a suppressant to the surface of the cooking equipment, inside the plenum of the hoods, and to the hood duct collar transition into the exhaust duct in the event of a fire. Downstream coverage of the grease duct varies by manufacturer and is typically up to 75 ft (22.9 m), at which point additional nozzles may be required for the downstream ducting. The governing standard for commercial kitchen fire suppression is UL 300 (2019). UL 300 is covered in NFPA 17A (2021a) and required in NFPA 96 (2021b).

The design of fire suppression in a grease-exhaust system is to be completed by a properly trained and certified agent of the equipment manufacturer and performed in accordance with the product's listing. Coverage methods include equipment-specific or flood systems, and detection methods are either mechanical or electronic. In any case, surface protection is specifically detailed in product-listing manuals and must be closely followed. The net requirements and methodology will determine the system's cost and size.

In addition to selecting the coverage method and detection type, the designer should understand the requirements when multiple hoods are specified for a building. In some cases, such as when hoods share a duct system or are installed end to end or back to back, each hood needs to simultaneously actuate the fire system in the event of a fire. Alternatively, in some cases, such as tenant kitchens, it is important for the fire suppression systems to be completely independent from one tenant to the next. Downstream, pollution control, heat recovery, or other grease-duct mounted accessories are included as a "common" hazard, and the fire suppression coverage must be designed in accordance with the locally adopted code. Pollution control or other duct accessories are required to have UL 300 (2019) or equivalent protection as required under NFPA 96 (2021b) and may also be required to discharge fire suppression when any of the other fire suppression systems discharge.

Fire systems are often connected to control sequences via a set of dry contacts such as a microswitch. As an example, one or more microswitches may trigger the following events:

- Activate the exhaust fan at 100% airflow
- Shut off the supply air local to the hood
- Initiate the gas shutoff valve and electrical shutoff responses for the cooking equipment
- Signal the building alarm system
- Signal any other connected systems or protections for a fire event, such as the building management system, if applicable

Discharge of the wet chemical occurs when the temperature in the hood is so high that it triggers the detection device in the hood, indicating that there is a fire, or when the manual pull station is activated to indicate a fire.

CKV Sequences and Testing, Adjusting, and Balancing (TAB) Requirements

Typically, the operation of the exhaust hood's fan and MAU are controlled in one of the following sequences:

- **Constant-volume CKV system sequence.** The exhaust fan's and makeup air fan's operations are interlocked, so if an appliance located under the hood is operated, both fans operate. This can be done using a wired connection between the fans and the appliance; however, it is typically accomplished by installing a temperature sensor in the hood—the fans are started when the temperature under the hood reaches a set point that indicates an appliance is being operated.
- **DCKV system sequence.** A DCKV system can be operated using one of the following methods. Consult with the specific DCKV manufacturer for more information.
    - Temperature sensors measure the hood temperature and operate the exhaust fan and makeup air fan at the desired condition.
    - Temperature sensors measure the hood temperature and room temperature and operate the exhaust fan and makeup air fan at the desired condition, depending on the difference in temperature between the two sensors.
    - In addition to using temperature sensors, the system uses an optical sensing device to detect if smoke from cooking is present.
    - Measure the cooking surface temperature using infrared (IR) technology and then control the fan's operation accordingly.
- **Dishwasher or other Type II hoods sequence.** Usually the fan's operation is controlled by a signal from the appliance (typically a dishwasher) or from a temperature sensor under the hood.

Any CKV sequences should work in harmony with the sequences of the surrounding space's HVAC sequences (if applicable).

For testing, adjusting, and balancing (TAB), the engineer should include the following in their specifications:

- The exact procedure for measuring the air being exhausted by the hood; this is critical, because different hood manufacturers have different air-measuring procedures.
- The velocity of any air that is being transferred from adjacent spaces should be measured with an appropriate measuring instrument. Additionally, the velocity of air at the perimeter

of the hood should also be measured to ensure that it does not exceed the criteria listed in this guide.

See chapter 8 for more details on system balancing and commissioning.

8

TAB and System Commissioning

OVERVIEW

Of all the activities covered in this guide, testing, adjusting, and balancing (TAB) and system commissioning (Cx) are the most critical, because these two activities ensure the commercial kitchen ventilation (CKV) system operates as per the engineer's design. Compared to the engineering and manufacturing of a new product, these steps for new construction are analogous to the "proof of concept" phase. Because TAB and system Cx are each subjects that are worthy of guidebooks themselves, only the critical details of these tasks as they relate to CKV systems are listed in the following sections.

Due to the fact that some TAB activities for CKV systems can be considered part of the Commissioning Process, this design guide combines these two activities into one chapter. Additionally, the Cx activities discussed in this chapter are written for the commissioning provider (CxP) to perform Cx; equipment suppliers may have their own specific Cx procedures for the start-up of their equipment.

Finally, this chapter was written to be used by both the CxP and the design engineer. The content can be used by the design engineer to create project specifications (TAB and Cx), and the CxP can use this content to create the Commissioning Plan. Some of the activities included in this chapter occur during the project's inception; however, most of the chapter's activities occur near the completion of construction, which is why this chapter is placed after the design chapters.

TESTING, ADJUSTING, AND BALANCING

TAB for CKV systems should be performed by a third-party contractor who has been certified by a recognized TAB certification agency. Examples of these certification agencies include National Environmental Balancing Bureau (NEBB) and Associated Air Balance Council (AABC). These agencies set minimum requirements for the experience, methods and processes, instrumentation, and reporting contents that are to be followed by contractors who perform TAB. It is recommended that the TAB contractor have previous experience with TAB for CKV systems. These requirements should be included in the project's TAB specifications.

While the proper operation of a CKV system is not as critical as that of a clean room or laboratory HVAC system, many of the design and operating approaches used in the these critical systems are also incorporated in CKV systems, including space pressurizations, proper air movement, and effluent capture. That being said, an unbalanced CKV system can result in an unhealthy and uncomfortable indoor environment, higher energy use, disruption to the kitchen operations, and even false fire alarms from improper capture and containment (C&C) of cooking smoke.

Additionally, many codes require that the CKV system be balanced, including *International Mechanical Code* (IMC) Section 507.6.1, which states, "[Exhaust hood] capture and containment test. The permit holder shall verify capture and containment performance of the exhaust system. This field test shall be conducted with all appliances under the hood at operating temperatures, with all sources of outdoor air providing makeup air for the hood operating and with all sources of recirculated air providing conditioning for the space in which the hood is operating located. Capture and containment shall be visually verified by observing smoke or steam produced by actual or simulated cooking, such as that provided by smoke generators" (ICC 2021d). Note that smoke bombs are not included in the list of acceptable simulated cooking products, because they emit smoke at a rate that is too rapid to represent realistic cooking.

Ensuring the hood captures and contains the cooking effluent is also covered in the "Commissioning" section of this chapter.

TAB Means and Methods

It should be noted that while the process of performing TAB can be considered both an art and a science (as can be said of the practice of engineering), certain measurements must be performed in the prescribed manner to prevent the possibility of erroneous measurements. This is especially true for the measurement of exhaust hood airflows, because hood manufacturers may have their own prescribed methods for measuring their hoods' airflows, and the method that works for one specific hood may not work for a different hood. If manufacturer recommendations are not provided, the TAB contractor performing CKV work should follow the recommendations of ASHRAE RP-1376 (Kuehn et al. 2010), *Method of Test to Evaluate Field Performance of Commercial Kitchen Ventilation Systems*.

To ensure this prescribed hood airflow measuring method is documented and followed, it is strongly recommended that the project's written specifications require that this documentation be submitted as part of the exhaust hood's submittal package for review by the engineer of record. This will ensure that the process has been documented and communicated to the project's stakeholders. Note that this requirement should also be included in the project's Commissioning Plan.

Examples of systems that are to be tested as part of the proper operation of a CKV system include the following:

- Type I exhaust hoods and their associated exhaust fans
- Type II exhaust hoods and their associated exhaust fans
- Any general exhaust fans that operate to maintain space pressurization
- Any HVAC systems that provide comfort heating or cooling to the space
- Any systems dedicated to providing or that provide part of the makeup air or replacement air that is necessary to maintain proper space pressurization of the CKV system—this could also include surrounding spaces, zones, or systems
- All transfer air from adjacent spaces—this should also include space pressurization measurements
- Any part-load or temporary/transient operational conditions, such as those that would be encountered in a demand-controlled kitchen ventilation (DCKV) system or during night operation for a 24-hour kitchen
- Verify that all occupied and unoccupied sequences of operations and controls are working as specified

Specific for the TAB for CKV systems, it is necessary for all HVAC equipment to be properly installed, started up, and made operational, including all items that impact the airtight sealing of the exhaust hood's ductwork, such as access panels.

It is critical that all systems associated with the CKV system's proper operation are tested and balanced at the same time. This is necessary because of the interoperability of the different systems (i.e., the exhaust system cannot be balanced unless the replacement and makeup air systems are operational).

In addition to performing TAB for new construction, TAB must also be performed for the entire CKV system a maximum of every five years of operation and when any of the following occur:

- New kitchen equipment or HVAC equipment is installed or equipment is replaced
- Modifications are made to the existing kitchen equipment cooking lineup that is located under an existing-to-remain exhaust hood
- CKV performance issues are observed (e.g., smoke loss from the exhaust hoods, high velocity drafts from adjacent spaces, etc.)

When TAB is not properly performed, or not performed at all, the following can occur:

- Drafts from adjoining spaces or from the outdoors
- Transfer of cooking odors to adjacent non-kitchen and dining areas
- Formation of condensation on diffusers or other cold surfaces
- Comfort issues in the kitchen and/or dining areas or spaces adjoining the areas served by the CKV system
- Exhaust hoods not being able to capture the cooking effluent
- Natural draft flues, such as those on domestic water heaters, not able to properly operate
- Excessive cycling of the heating or cooling system
- Indoor environmental quality (IEQ) issues resulting from cooking effluent not being removed from the breathing zone
- Excessive energy usage
- Excessive operation and maintenance (O&M) manpower efforts and costs
- Infiltration of moisture into building envelope

Note that almost all CKV operational system issues can be corrected by executing proper TAB.

When balancing existing kitchens where no design airflow values are available (such as a remodel), it is recommended to research any applicable codes followed by the local authority having jurisdiction (AHJ) to determine if they have any specific requirements. If necessary, Table 8.1 in Chapter 34 of *ASHRAE Handbook—HVAC Applications* (2019) and Table 8.2 in ANSI/ASHRAE Standard 154 (2022d) can be referenced as a guideline of typical hood exhaust airflow rates for Type I hoods and minimum airflow rates for Type II hoods.

TAB Process and Approach

The following list contains specific items and procedures that must be followed or included when performing TAB for any CKV system. Note that the sections are in this list are organized into bullet points to allow easier incorporation into project specifications and site observation checklists.

- Follow one or more of the listed referenced standards or guidance:
    - *Procedural Standard for Testing, Adjusting and Balancing of Environmental Systems* (NEBB 2019)
    - *National Standards for Total System Balance* (AABC 2016)
    - *ASHRAE Handbook—HVAC Applications* (2019), Chapter 39, "Testing, Adjusting, and Balancing"

- The following should be evaluated:
  - The exhaust hood manufacturer's air measurement recommendations and examples. This may include specific procedures when measuring hoods with multiple supply air or exhaust air duct connections.
  - The general approach(es) and order when performing any TAB.
  - The k-factors of all the air supply and exhaust devices (e.g., terminal units), especially the exhaust hoods.
  - The experience and knowledge gained from previous CKV TAB work, including recommended procedures from the instrument's equipment manufacturers and the referenced TAB procedural standards.
- Accurate airflow measurements are dependent on selecting the right instruments for the application, maintaining calibration of those instruments, and using those instruments correctly. Commonly used TAB tools and instruments include the following:
  - Manometer
  - Pitot tube
  - Velocity grid (Figure 8.1)
  - Rotating vane anemometer (Figure 8.2)
  - Airflow balometer/airflow capture hood
- The recommended CKV TAB processes—including critical steps, methods, exceptions, and variations—and when to apply them are as follows:
  - Confirm that all CKV equipment is operational and ready for balancing (i.e., equipment has been started up, fans are rotating in the correct direction, belts are tight, filters are clean, access panels are secure, etc.). This includes all related exhaust fans, makeup air fans, replacement air systems, and space heating/cooling systems.
  - Ensure all balancing dampers are fully opened and outdoor air dampers on any replacement air systems are at least open to minimum position.
  - Set controls to operate fans continuously and at maximum flow.
  - Measure all initial CKV airflows. This includes all related fans, exhaust hoods, outdoor airflow quantity, supply air diffusers, return air grilles, exhaust air grilles, etc.
  - Compare all initial airflow measurements to design requirements and determine what adjustments are needed.
  - Adjust all CKV airflows to achieve the designed airflows. This is accomplished by changing fan speeds and sequentially adjusting balancing dampers. Multiple adjusting iterations are typically required, because the systems operate interdependently (i.e., as one piece of equipment is adjusted the others can be affected).
    - Due to measuring instrument accuracies and the effects of field conditions, the designed airflows should include a specified design tolerance. The typical tolerance is ±10%, meaning that the airflows are to be balanced to within a range of 90% to 110% of the designed airflow value. Tighter tolerances are possible but can greatly increase the time required to perform the TAB work. That being said, the airflows must be adjusted so that the CKV design's intended pressurization and air movement are obtained. For measuring hoods/fans, larger airflows should be considered because they allow for tighter tolerances.
    - An efficiently balanced air distribution system will have at least one balancing damper fully opened so the fan airflow is not restricted, optimizing fan performance.
    - If system deficiencies prevent the design airflow from being obtained, immediately notify the owner's representative and/or appropriate contractor, engineer of record, and manufacturer.
  - If the CKV system has a demand-controlled ventilation system, set the controls for minimum flow, remeasure airflow, and calibrate the controls as needed to achieve the

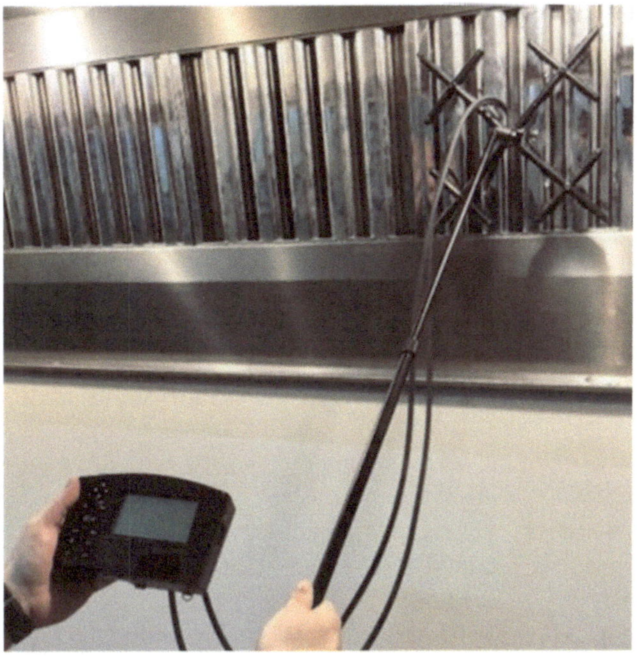

Figure 8.1 Example of a velocity grid being used to measure airflow through the grease filters of an exhaust hood.

Courtesy of Melink Corporation

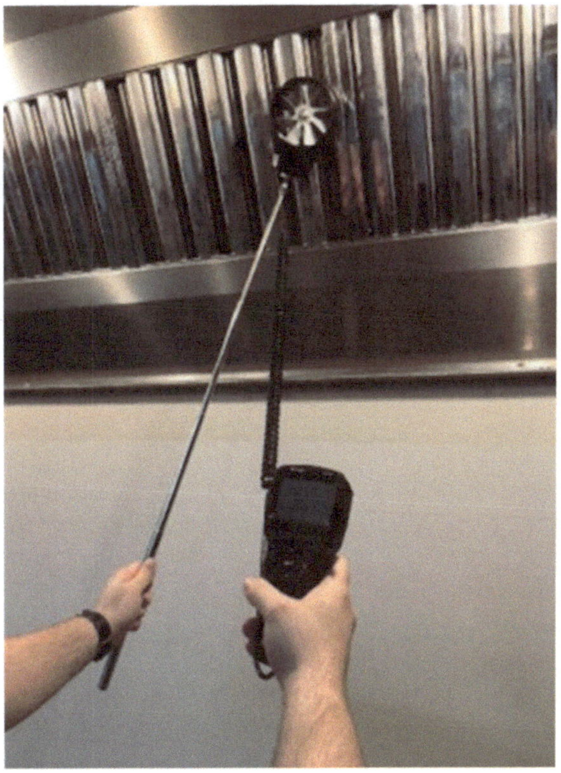

Figure 8.2 Example a rotating vane anemometer being used to measure airflow through the grease filters of an exhaust hood.

Courtesy of Melink Corporation

design minimum airflow. Also, while performing these measurements, be sure that any associated HVAC systems (e.g., replacement or makeup air equipment) are also operated at their minimums to maintain proper pressurization.
    - Look for any high-velocity drafts near the exhaust hoods.
    - Observe exhaust hood performance with all CKV systems operating.
    - Ensure cooking appliances operate at cooking temperatures.
    - Ensure actual cooking is taking place to verify C&C of the cooking effluent.
    - If C&C is not achieved, look for sources of disruption and make adjustments to the CKV system airflows to correct it.
    - Measure space pressurization and compare the valves to design requirements. Make final adjustments to correct any pressure issues.
    - Record final balanced airflows, control settings, and equipment operating data (e.g., fan revolutions per minute, motor amps, static pressures, damper positions, etc.) to serve as future reference points.
- The following should also be observed and verified for proper equipment installation as well as proper CKV system operation, because these items can impact obtaining completion of proper TAB. Note that these items are also typically performed by the CxP as part of their prefunctional checks:
    - Verify that the hood overhang for the cooking appliances matches the hood submittal drawings. If necessary, correct by pushing all appliances as far away from the open sides of the hood as possible. Where sufficient overhang cannot be achieved, as a possible solution side panels could be installed on the hood to enclose the appliances. Note that the determination regarding the addition of side panels needs to be made by the engineer of record in coordination with the hood supplier.
    - Any high-velocity drafts near the exhaust hood, such as those from supply air diffusers, fans, transfer air outlets, etc., will impact the hood's ability to capture and contain cooking effluent. Any observable drafts should be noted in the TAB report, and if any corrective measures are made to reduce or eliminate them, those measures should also be documented in the TAB report.
    - Similar to noting drafts, also note if there are any nearby return or exhaust grilles or hoods affecting the operation of the kitchen hood being balanced.
    - Note if any exhaust hoods are missing filters, have incorrect filters installed, or have filters that are not fitted tightly (i.e., there are gaps between the filters). Any of these deficiencies will have an impact on the air quantity exhausted by the hood as well as the ability to properly balance the CKV system. Correct any deficiencies before taking measurements.
    - Ensure that all of the exhaust duct clean-out panels have been installed and tightly secured.
    - Ensure there are no issues with the exhaust fan's proper operation (e.g., backward fan rotation, loose belt, the fan is not secured tightly to the roof curb, there are improper duct transitions, etc.).

Post-Occupancy Air Balance Observations and Verifications

The following list contains items that have been observed to occur in CKV systems after the kitchens have been occupied and operated and that may impact the system's air balance and performance. Some of these items occur during the normal operation of a CKV system, while others are the result of improper or deficient O&M. As part of the CKV's O&M plan, it is recommended that an inspection of the kitchen be conducted a minimum of once every three months to identify and correct any of the following:

- The cooking appliances have been moved too close to the edge of the hood or even out from under the hood's canopy.
- The cooking appliance lineup has changed from the original design. This can include the appliance type, cooking duty, quantity, and even location.
- The exhaust hood's grease filters are damaged, missing, or not fitted tightly (i.e., there are gaps between filters).
- Lack of routine proper cleaning has resulted in grease building up on the following:
    - Exhaust duct walls
    - Exhaust fan's wheel
    - Hoods filters
- The exhaust duct's clean-out panels are missing or not secured tightly, allowing air to leak into the duct.
- The exhaust fan's belt is loose or broken.
- The outdoor air intake dampers are closed, due to either a malfunction or improper maintenance.
- The filters and/or screens that serve makeup air plenums are loaded with debris.

Additionally, even under proper use, a CKV system should be rebalanced when any one of the following occurs:

- Smoke or steam is observed escaping from the exhaust hoods.
- Grease is observed to be accumulating on the ceiling around the perimeter of the exhaust hoods.
- Grease is observed to be accumulating at return grilles/ducts.
- Cooking odors are migrating to adjacent areas.
- There are noticeable high-velocity drafts from adjacent spaces and/or through any pass-through windows (or openings) to other adjacent interior spaces.
- Air is being drawn into the building when any exterior doors and/or exterior windows are opened.
- There is an unusually high space humidity, above 65% rh (per the recommendations from ASHRAE RP-1469 [Stoops et al. 2013]) in any of the CKV areas during cooling operation when the outdoor air is humid (i.e., the outdoor dew point is above 65°F [18°C]).
- There are uncomfortably high or low space temperatures that cannot be properly controlled by the CKV system's heating and cooling cycles.
- The spaces served by the CKV system have been remodeled.

As part of the O&M plan, the CKV system should be tested, adjusted, and balanced once every five years (or more frequently), because the following can occur:

- Normal changes and variations to the HVAC equipment's operation.
- Worn-out equipment components and devices due to wear and tear from typical operations.

As part of this plan, verification for being within operational tolerance should be performed once every five years (or more frequently). In addition to the above-listed circumstances, the outdoor air quantities should be measured and adjusted as necessary every two years, because of the challenging operating conditions under which outside air dampers typically operate.

For additional information, refer to the *ASHRAE Handbook—HVAC Applications*, chapter 34, "Recommended Frequency of Maintenance" section (2019).

Commissioning (Cx)

For this section, the expectation is that the reader has experience in the Commissioning Process as described in ANSI/ASHRAE/IES Standard 202 (ASHRAE 2018). The items covered in this section are topics that are specific to commissioning CKV systems. They include everything from the project conception to construction and turnover to ownership as well as recommissioning and retrocommissioning activities.

Specifically, this section focuses on CKV-specific elements of the following Cx items:

- The Owner's Project Requirements (OPR) and the Basis of Design (BOD)
- The design review and what to focus on
- System verification (i.e., prefunctional checks)
- Functional performance testing
- Training
- Post-occupancy observations/verifications

These items should be included as necessary to ensure proper CKV system operation and are required to ensure the project achieves its goals, including obtaining any necessary certifications, such as LEED® (USGBC 2023), WELL (IWBI n.d.), or others.

Owner's Project Requirements (OPR)

As is typical for any project following the Commissioning Process, during this phase the focus should be on discussions with the owner or their representative to determine their goals for the project as they relate to the kitchen and adjoining areas. Determining these goals involves discussions with the food service consultant—whose expertise lies in designing the kitchen's layout and selecting and locating the appliances—along with discussions that should include the project's architect, mechanical engineer, general contractor, and general contractor's mechanical, electrical, and plumbing (MEP) subcontractors. Additionally, any of the owner's food service experts (e.g., facility and operations managers) and stakeholders should also be included.

This coordination with all the disciplines and trades, and the associated continuous alignment of the project's details to achieve the project's goals, should continue throughout the entire life the project. Focusing on coordination between the different design disciplines is especially critical for a CKV system, because the decisions made by one discipline will greatly impact the requirements and design decisions made by the others. As an example, hood selection will impact exhaust airflow rates, which will in turn impact the entire CKV design.

The OPR's goals may include any or all of the following:

- Energy-efficiency targets
- Certifications (e.g., LEED)
- Cost or budget requirements
- Kitchen IEQ, especially the target operating temperature and humidity (e.g., deciding what is too hot)
- Location of kitchen, dining area, and any applicable adjoining spaces whose operation or IEQ may be impacted by the CKV system
- Hours and days of operation (critical if the kitchen is located in a larger building, such as a cafeteria in an office building)
- Number and types of cooking platforms (e.g., charbroilers and other high-temperature cooking appliances)
- Odor control to adjacent spaces and/or the outdoors (e.g., are the neighbors a concern?)
- Sources of makeup air/replacement air for the kitchen hood's exhaust

- Maintenance expectations—these are a very critical and must be defined by the owner and complied with by all stakeholders. The key question is, "Will the CKV system be maintained by a trained engineering staff, an outside maintenance company with expertise in CKV systems, or by the kitchen managers and their staff (as is typical for many standalone restaurants)?"
- Planned future modifications (if applicable)

While the CxP may be the person who facilitates these discussions, by no means are they the only person who should be leading these discussions or acting as the expert on these subjects. The goal when creating the OPR is to ensure these CKV-related discussions do occur and that their outcomes are documented in the OPR.

Basis of Design (BOD)

In addition to a typical BOD, the CKV BOD should specifically list/detail the following:

- Each specific cooking platform, including the following:
  - The type of cooking appliance (e.g., fryer, griddle, etc.) and food being cooked
  - The type of hood that will serve the appliance (i.e., canopy, island, backshelf)
  - Any modifications to the hood(s) such as adding side panels, integral airflow devices, etc.
  - If DCKV controls will be incorporated
- The kitchen hood exhaust system including the following:
  - Pre-fab or field-fabricated exhaust duct
  - If any pollution control devices (e.g., scrubbers) will be used (if yes, the specific devices must be listed)
  - The location of the exhaust system's mechanical equipment (e.g., fans, scrubbers, etc.)
  - Any special architectural considerations to provide access for maintenance of the exhaust system's mechanical equipment and the duct cleaning access panels
- The method of providing and conditioning (controlling the temperature and humidity) the replacement air for the kitchen hood's exhaust system, such as the following:
  - Direct from outdoors
  - Transfer from adjacent spaces
  - A combination of the above
- Any necessary space pressurization requirements (list the adjacent spaces)
- The IEQ of the kitchen and dining areas, including the following:
  - Design temperatures for all spaces served by the CKV system
  - Design relative humidity
  - Control and/or mitigation of any sources of indoor pollutants (e.g., particulates and volatile organic compounds [VOCs]) that are a byproduct of the kitchen's cooking operations, including unhooded appliances, and the effluent produced by food after it is removed from a cooking process (if you can smell it, it is producing byproducts)
- The type of kitchen hood fire suppression system that will be used and the locations and equipment that will be served by it
- A high-level description of any CKV control sequences of operation, including interlocks and, if applicable, DCKV systems, and the locations of all components
- Accommodations for future modifications, especially regarding the cooking platforms

Commissioning Plan

The Commissioning Plan for a CKV system includes the typical Cx activities of the HVAC equipment and systems associated with a CKV system (such as fans, air-handling units, etc.) and should also include the following activities that are unique to CKV systems:

- Performing a prefunctional checklist visit to verify proper installation of the following:
  - The hoods and accompanying accessories along with any appliances they serve
  - The locations of all HVAC equipment and devices, with particular attention paid to the locations and type of air terminals (return and supply)
  - Proper access for required equipment and system maintenance
  - Exhaust fans
- Verifying all interlocks between the kitchen appliances and CKV system
- Ensuring the hood fire suppression system has been properly started up and commissioned by the hood fire suppression contractor (note that this system is separate from the building's fire sprinkler system)
- Verifying that the hood fire suppression system's activation has been properly interlocked with any applicable CKV HVAC systems
- Observing and verifying proper air movement of the CKV system, including the following:
  - Air movement adjacent to the hood(s)
  - The pressurization of replacement air from any adjacent spaces
- Observing and verifying the C&C of all exhaust hoods
- If applicable, verifying the proper operation of any DCKV systems

Design Review

When performing a design review of CKV systems, the following are some items that should be specifically reviewed to understand the design so it can be commissioned and to ensure the design meets the expectations and requirements of the OPR and BOD:

- The appliance hood drawings provided by the hood supplier. These drawings will provide details of the hood(s), including the following:
  - Hood type
  - Exhaust airflow
  - Any supply or replacement airflows (if the hood includes these options)
  - Any side panels (if the hood includes these options)
- The overall CKV system design, including the following:
  - Source(s) of replacement air
  - Any space pressurization or transfer air requirements
  - All exhaust, including hood and general space exhaust
  - The location and type of any supply or return/exhaust air devices to determine if they might impact hood C&C
- Any CKV system controls, including the following:
  - Control and operation of any replacement air sources
  - Any associated exhaust air controls and interlocks, including the those of the cooking appliances
  - If applicable, the CKV system's DCKV controls, including what parameters cause the exhaust airflow rates and makeup air rates to modulate and the corresponding modulation rates of both (note that some systems have only two speeds while others have variable speeds)

- The hood fire suppression systems, including the following:
  - The operation of the CKV system as well as the operation of any associated HVAC systems to understand how they interact with the operation of the fire suppression system
  - How any associated cooking appliance's fuel source is shut off
  - Simultaneous operations of systems (i.e., common hazard)
- If applicable, any kitchen exhaust pollution control devices

For additional design details, refer to chapters 3 and 4 of this guide.

Prefunctional Checks

The following list contains the equipment and components that are unique to CKV systems and that should be observed during construction and prior to functional testing for proper installation and compliance with the project's goals and requirements:

- The exhaust hood and all its applicable accessories. Verification may include the following:
  - The intended maximum operating temperature (depends on the appliances it serves)
  - That grease filters are properly installed
  - Side panels, where specified
  - If applicable, DCKV control sensors, which can be located within the hood, in the exhaust duct just above the hood, or in the kitchen space
  - Fire suppression systems
  - If applicable, local manual controls, such as ON/OFF switches and overrides
- The cooking appliances under the hood, including the appliance types (fryer, charbroiler, etc.) and their exact locations, which should match the hood submittal drawings. Also verify proper hood overhangs, which are based on the hood submittal drawings. Note that even the act of relocating appliances under a hood, such as swapping the locations of a griddle and charbroiler, will impact the hood's ability to capture and contain.
- The exhaust duct installation including:
  - Listed assemblies for listed prefabricated duct systems.
  - Welded liquid-tight construction for field-fabricated exhaust ducts for Type I hoods. Typical HVAC duct construction, including folded seams, is not acceptable for Type I kitchen exhaust hoods but is acceptable for Type II kitchen exhaust hoods.
  - Listed duct access panels in locations as mandated by code and perceived reasonable access for accessibility. Refer to chapter 5 for details and rationale.
- The exhaust fan, ensuring that it is listed for its intended cooking application and temperature and includes access panels for periodic cleaning of the fan wheel and housing, which occurs when the associated exhaust duct is cleaned.
- Pollution control devices, ensuring access is provided to replace any filters or other replaceable components. Additionally, if technologies such as electrostatic precipitators or ultraviolet-C (UVC) systems are used, verify that they have been properly started up by a qualified technician and that a start-up report has been made available.
- The DCKV system has been started up by a qualified technician and a start up report has been made available.
- Replacement air systems have all been installed as per the design and are installed in their prescribed locations. This includes verifying the locations of all supply and return/exhaust air terminals that are part of the CKV system. It also includes those items necessary to maintain proper space pressurization and may also include components of separate HVAC systems.
- Diffusers. Diffuser types can be changed during construction for aesthetic or supply (e.g., lead time) reasons. These types of changes/substitutions can have a negative impact on a

CKV system's performance. It is always best to confirm any changes with the engineer of record. Refer to chapter 4 of this design guide for verification of diffuser selection.
- TAB has been performed and all issues and deficiencies have been resolved. TAB is the foundation of CKV commissioning.

Functional Performance Testing

One of the most critical Cx tasks is conducting functional performance testing. These are tests in which system operation is evaluated for the engineer's design intent. More specifically, testing of the system's sequence of operations is performed to ensure the system operates according to the prescribed sequence.

This section discusses elements that are unique to CKV systems and that must be tested or observed to ensure the CKV system operates as intended. It is expected that the reader of this section has experience testing regularly encountered HVAC equipment such as air-handling units (AHUs), makeup air units (MAUs), variable-air-volume (VAV) systems, and exhaust fans.

In the following subsections, a suggested corresponding functional performance test script follows each area of focus. These areas of focus are items that should be tested prior to owner turnover. Additionally, some local AHJs may have their own functional performance test requirements that are similar to those listed below. As an example, there are functional performance test requirements such as those listed in California Energy Commission's (CEC; 2019) *Nonresidential Compliance Manual*, Section 13.29, NA7.11.1 Commercial Kitchen Exhaust. The commercial kitchen exhaust acceptance tests listed in that manual are a recommended starting point for creating functional performance test scripts.

Appliance/hood/replacement air interlock. Verify that the interlocks between the appliance's operation, hood exhaust fan, and replacement air system(s) operate so they all turn on and off as prescribed. These interlocks can be initiated using or more of the following methods:

- Relays (for wired interlocks)
- Building automation system (BAS) controls
- Hood/room temperature sensor differential

Be sure to observe, verify, and document any minimum runtimes. (Sometimes, a 30-minute minimum system runtime is programmed to ensure the ON/OFF sequence does not repeatedly cycle the affected system's operations.)

Suggested test script:

1. Turn on the cooking appliance.
2. Observe whether the hood exhaust fan operates.
   a. If a hood/room differential temperature sensor operation is used, document the amount of time between appliance's ON operation and fan's start operation.
   b. If an electrical interlock is used between the appliance and the hood exhaust fan, test each appliance individually to ensure that when it is turned on, the hood fan operates.
3. Ensure the replacement air system (AHU, MAU, etc.) operates when the hood exhaust fan operates. This includes the operation of any fans that are integral to the hood.
4. Turn off the cooking appliance.
   a. If a hood/room differential temperature sensor operation is used, document the amount of time between the appliance's OFF operation and the fan's shutdown operation. (Note that this is typically 30 minutes but will be based on the hood manufacturer's controls.)
   b. If an electrical interlock is used between the appliance and the hood exhaust fan, test each appliance individually to ensure that when it is turned on, the hood fan operates.

3. Ensure the replacement air system (AHU, MAU, etc.) stops operating when the hood exhaust fan is shut off.

Cooking fire suppression system/CKV system interlock. Verify that the interlocks between the cooking appliance's fire suppression system and the CKV system operate as prescribed. The sequence of operation of these interlocks is mandated by the local AHJ and/or the local code. (Note that this specific test is not to verify the proper operation of the suppression system itself—this is typically performed by the fire suppression system contractor and witnessed by the AHJ—but is to verify the suppression system's operational interlock with the CKV system.)

In large systems, one or more of the following may be present and must be considered:

- Multiple hoods may be served by one exhaust fan. Typically, if any hood's fire suppression system discharges, the operation of the entire CKV system will be impacted.
- If the system contains any pollution control units (PCUs), those units will likely have their own fire suppression system that is interlocked with the hood's fire suppression system.

Suggested test script:

Note: This test requires the assistance of the fire suppression contractor, who will activate their cooking fire suppression system **without** it discharging the fire suppression agent.

1. Turn on the cooking appliance, hood exhaust fan, and replacement air system to simulate typical cooking operations.
2. The fire suppression contractor will activate the suppression system in a manner that does not cause the system to discharge the suppression agent.
3. Observe the operation of the following systems:
   a. The fuel source to the appliance shall be disconnected. This can be accomplished in one of two ways:
      i. For electrical appliances, the electrical power to the appliance is disconnected, typically using a shunt-trip breaker.
      ii. For gas appliances, there will be an emergency automatic gas shutoff valve.
   b. Typically, the source of supply/makeup air will be shut off. (Note that this depends on the local code.)
   c. Typically, the hood exhaust fan continues to operate at full speed; the intent is to remove smoke from the occupied space.

CKV replacement air and pressurization. Verify that the exhaust hood will be able to properly capture and contain effluent, including observation of the replacement air systems, air movement around the hood, air movement into and through the CKV space(s), and proper space pressurization that follows the design intent. Observe the system to ensure that there are no system negative effects, such as the following:

- High-velocity air supplied near any hoods (e.g., comfort fans)
- 75 fpm (0.381 m/s) maximum air speed at hood edge (these numbers should be confirmed on the TAB report)
- 75 fpm (0.381 m/s) maximum transfer air velocity from adjacent spaces (these numbers should be confirmed on the TAB report)

The replacement air may be provided by separate HVAC systems.

Suggested test script:

With the CKV system in normal operational mode, complete the following:

1. Observe the air movement within the CKV space(s) to determine whether any air flowing near the hood will impact its ability to capture and contain effluent.
2. Observe the adjacent spaces and the associated pressurization and air movement between them to ensure the air movement and pressurization are per the engineer's design intent.

A smoke pencil is very useful for observing air movement and determining pressurization.

DCKV system operation (if applicable). Verify that the demand control system operates as intended and that the CKV system responds accordingly. Each manufacturer's demand control system operates in its own prescribed manner, so any test scripts must be written for each project's specific system. Typically, these systems incorporate one or more of the following, with the goal of automatically increasing or decreasing the exhaust air and makeup air quantities based on cooking activities:

- A measurement of the temperature differential between the inside of the hood and the adjacent room. If the temperature difference is small, the fan operates at a reduced speed; if the difference is large, the fan operates at full speed.
- A light beam measurement across the inlet of the hood to detect any smoke from cooking. If the beam is blocked, the exhaust fan's speed increases.
- A thermal measuring device, similar to that of a thermal measuring gun or camera, which is located in the hood and measures the temperature of the cooking appliances' surfaces and adjusts the exhaust fan speed based on the observed cooking operations.

DCKV systems must be started up by a qualified manufacturer's technician, and a start-up report of the hood's DCKV controls must be submitted to the CxP as part of the Cx of these systems. Additionally, the assistance of a manufacturer's technician may be required to perform these tests. Note that some systems have only two speeds, while others have variable speeds.

Suggested test script:

1. Turn on the cooking appliance, hood exhaust fan, and replacement air system to simulate part-load cooking operations that would occur at the start of a cooking operations work shift (e.g., when the cooks arrive to begin preparing breakfast or lunch).
2. Observe the exhaust air fan speed for compliance with DCKV part-load cooking specifications.
3. Test the different ventilation operational conditions by performing the following (as required):
   a Either turn appliances on to high or use a heat gun to heat the hood's temperature probe to test the temperature differential sequence.
   b Physically block the light beam sensor to simulate smoke-filled cooking conditions.
   c Turn the appliances on to high for the thermal imaging sensor.
4. For each of the above test steps, observe the makeup air's system operation to ensure parallel operation with cooking exhaust fan operation (e.g., the makeup air quantity increases and decreases at the same rate as the exhaust air quantity changes).
5. For each different exhaust and makeup air operational condition, observe the adjacent spaces and the associated pressurization and air movement between them to ensure the air movement and pressurization is not affected by the change in exhaust air and replacement air quantities.

Post-Occupancy Testing (e.g., CKV Seasonal Testing)

While functional performance testing can be used to commission most of the CKV system, verification of the entire system's proper operation should be performed during normal cooking operations. Because many jurisdictions do not allow cooking operations to be performed until after the certificate of occupancy has been awarded, some of the testing must be performed after the space is occupied and functional. Such tests are unobtrusive and can be performed mostly by observing operations and monitoring the system's operations by analyzing trend data. (This follow-up site observation visit and the associated testing can be compared to a Cx follow-up seasonal site visit for a heating or cooling system and should be planned for when creating the proposal for Cx services and the subsequent Commissioning Plan.)

These tests should include the following actions:

- Witness the exhaust hoods' C&C under full-load cooking conditions or, at a minimum, the cooking appliances at the ends of the hood operating under full-load conditions. All the smoke produced by the cooking must be captured—**none of the smoke should spill out from under the hood**. If complete C&C is not obtained, a mitigation measure or plan must be developed to obtain complete C&C. Measures can include:
  - Identifying and eliminating air currents or drafts occurring near the hood
  - Adding side panels to the hood to increase the hood's C&C performance
  - Increasing the exhaust air quantity (which will also require increasing the makeup air quantity to maintain proper air movement and pressurization)
- Observe pressurization and air movement within the CKV spaces and adjoining spaces to ensure they comply with the engineer's design intent.
- Observe the CKV spaces for any thermal comfort issues or performance items that are impacting the space's IEQ. These may include makeup air set points, improper hood C&C, CKV air distribution, etc.
- Observe the cooking appliances' locations to ensure the appliances have not been moved from their locations as shown on the hood submittal, and if they have been moved, note it in the site visit report.
- For a DCKV system, also conduct an analysis of operational trend data to ensure the demand controls have been properly set up by the manufacturer during start-up and that the overall CKV system is operating properly. This includes operating the CKV system at slow speeds during idle cooking periods and at high speeds during full-load cooking periods.

Monitoring-Based Commissioning Plan for CKV Systems

If monitoring-based Cx activities are performed after post-occupancy testing has been completed, the following trend data should also be observed:

- Verify energy target goals.
- Verify proper system operation. This is especially critical for CKDCV systems.
  - Verify IEQ is maintained.
  - Verify proper appliance/CKV system ON/OFF operation is occurring.
- Identify areas and items whose operations and control sequences may need to be adjusted based on actual system operations.
- Verify that scheduled maintenance is occurring.

Some specific items that should be monitored include the following:

- Exhaust fan speed and exhaust airflows
- Replacement air quantities
- CKV space temperature and humidity
- CKV space particulates concentration measurements
- CKV space pressurization measurements

Retrocommissioning and Recommissioning

Retrocommissioning or recommissioning can be considered on an owner-determined basis to ensure proper CKV system operation throughout its operable lifetime. This design guide recommends that retrocommissioning should be considered at five-year intervals, at a minimum, during the CKV system's lifetime. Recommissioning should also be performed when any changes are made to the CKV system or to the spaces it serves. This includes any changes to the cooking appliance lineup.

A

Codes and Standards

Introduction

Commercial kitchen ventilation (CKV) installations for the North American marketplace are governed by a number of codes and standards from institutions including ASHRAE, ASTM International, International Association of Plumbing and Mechanical Officials (IAPMO), International Code Council (ICC), National Fire Protection Association (NFPA), and Underwriters Laboratories (UL). Note that the IAPMO *Uniform Mechanical Code* closely follows the NFPA 96 standard. Design engineers are encouraged to consult the authority having jurisdiction (AHJ) to clarify state and local requirements for a given project, as adoption varies widely.

There are other standards and some non-mandatory guidance available for designers as well. For example, ASTM International performance standards enable laboratory data to be generated for comparison purposes. These data may not be reflective of field conditions but can aid the designer in the selection process. ASHRAE Technical Committee (TC) 5.10 authors the "Kitchen Ventilation" chapter in *ASHRAE Handbook—HVAC Applications* (2019), which contains a wide range of information on the design process for CKV applications. The "Nonresidential Cooling and Heating Load Calculations" chapter in *ASHRAE Handbook—Fundamentals* (2021b) contains tabular data that can be used to estimate heat gain to the space for several commercial kitchen appliances.

Relevant standards' purpose and scope texts (where available) are presented here for review by the design engineer. Full citations for all these sources are included at the end of this chapter.

Note that everything that follows is reproduced verbatim from each code or standard.

ASHRAE Standards and Non-Mandatory Guidance

ANSI/ASHRAE/IES Standard 90.1, *Energy Standard for Buildings Except Low-Rise Residential Buildings*

1. Purpose
   1.1 To establish the minimum *energy efficiency* requirements of *buildings* other than *low-rise residential buildings* for
   a. design, *construction*, and a plan for operation and maintenance; and
   b. utilization of *on-site renewable energy* resources.
2. Scope
   2.1 This standard provides
   a. minimum *energy*-efficient requirements for the design and *construction*, and a plan for operation and maintenance of,

1. new *buildings* and their *systems*,
2. new portions of *buildings* and their *systems*,
3. new *systems* and *equipment* specifically identified in this standard that are part of a *site*,
4. new *systems* and *equipment* in *existing buildings*, and
5. new *equipment* or *building systems* specifically identified in the standard that are part of *process applications*

and

b. criteria for determining compliance with these requirements.

2.2 The provisions of this standard do not apply to

a. single-family houses and related incidental *structures*, multifamily *structures* of three *stories* or fewer above *grade*, manufactured houses (mobile homes), and manufactured houses (modular) or
b. *buildings* that use neither electricity nor *fossil fuel*.

2.3 Where specifically noted in this standard, certain other *buildings* or elements of *buildings* or *sites* shall be exempt.

2.4 This standard shall not be used to circumvent any safety, health, or environmental requirements.

ANSI/ASHRAE Standard 62.1, *Ventilation and Acceptable Indoor Air Quality*

1. Purpose

1.1 The purpose of this standard is to specify minimum ventilation rates and other measures intended to provide indoor air quality that is acceptable to human occupants and that minimizes adverse health effects.

1.2 This standard is intended for regulatory application to new buildings, additions to existing buildings, and those changes to existing buildings that are identified in the body of the standard.

1.3 This standard is intended to be used to guide the improvement of IAQ in existing buildings.

2. Scope

2.1 This standard applies to spaces intended for human occupancy within buildings except those within dwelling units in residential occupancies in which occupants are nontransient.

2.2 This standard defines requirements for ventilation and air-cleaning-system design, installation, commissioning, and operation and maintenance.

2.3 In addition to ventilation, this standard contains requirements related to certain contaminants and contaminant sources, including outdoor air, construction processes, moisture, and biological growth.

2.4 This standard does not prescribe specific ventilation rate requirements for the following:

a. Spaces that contain smoking or that do not meet the requirements in the standard for separation from spaces that contain smoking
b. Patient care areas not listed in this standard
c. Laboratories with hazardous materials

ANSI/ASHRAE Standard 154, *Ventilation for Commercial Cooking Operations*

1. Purpose

   The purpose of this standard is to provide design criteria for acceptable performance in commercial cooking ventilation systems.

2. Scope

2.1 This standard covers
  a. kitchen hoods,
  b. exhaust systems, and
  c. replacement air systems.

2.2 This standard shall not be used to circumvent any safety, health, or environmental requirements.

ASHRAE Handbook—Fundamentals, Chapter 18, "Nonresidential Cooling and Heating Load Calculations"

All cooking appliances should be considered when determining cooling loads for a building. This chapter provides guidance on nonresidential cooling and heating loads including that from unhooded electric appliances, hooded electric and gas appliances, ware-washing applications, and meals. Tabular data are presented for a wide range of tested appliances detailing radiation and usage factors that can be applied for generic appliances.

ASHRAE Handbook—HVAC Applications, Chapter 34, "Kitchen Ventilation"

This chapter focuses primarily on commercial kitchen ventilation (CKV) systems in restaurants and institutional food service facilities and incorporates the research and experience (including all steps of the design process) amassed by TC 5.10 over the past two decades. Although a brief section on residential kitchen ventilation has been retained, only minor updates have been incorporated over the history of this chapter. Given ongoing debate on residential range hood performance, TC 5.10 anticipates sponsoring research to provide a basis for enhanced design of residential kitchen ventilation (RKV) systems.

ASTM International Standards

ASTM E2336, *Standard Test Methods for Fire Resistive Grease Duct Enclosure Systems*

Purpose

The general purpose of this standard is to determine the fire resistance of assemblies which cover or enclose grease duct. Usually, the grease duct is unlisted field fabricated and usually the enclosures are flexible foil-faced grease duct wraps.

1. Scope

NOTE 1: The majority of this standard is based on the Model Building Code Evaluation Service Acceptance Criteria titled *ACCEPTANCE CRITERIA FOR GREASE DUCT ENCLOSURE ASSEMBLIES, AC101*, which was created in 1994. Numerous design listings and labeled materials exist based on the provisions of this standard.

1.1 These test methods evaluate the enclosure materials and the grease duct enclosure systems using the following test methods: noncombustibility, fire resistance, durability, internal fire, and fire-engulfment with a through-penetration fire stop.

1.2 These test methods prescribe a standardized fire exposure for comparing the test results of the enclosure materials and grease duct enclosure systems. The results of these tests are one factor in assessing predicted fire performance of grease duct enclosure systems. Using these test results to predict the performance of actual grease duct enclosure systems requires the evaluation of test conditions.

1.3 The values stated in inch-pound units are to be regarded as standard. The values given in parentheses are mathematical conversions to SI units that are provided for information only and are not considered standard.

1.4 The text of these test methods references notes and footnotes which provide explanatory material. These notes and footnotes (excluding those in tables and figures) shall not be considered as requirements of the fire test response standard.

1.5 *These test methods are used to measure and describe the response of materials, products, or assemblies to heat and flame under controlled conditions, but does not by itself incorporate all factors required for fire hazard or fire risk assessment of the materials, products, or assemblies under actual fire conditions.*

1.6 *These test methods do not purport to address all of the safety concerns, if any, associated with its use. It is the responsibility of the user of these test methods to establish appropriate safety and health practices and to determine the applicability of regulatory limitations prior to use.*

1.7 *This international standard was developed in accordance with internationally recognized principles on standardization established in the Decision on Principles for the Development of International Standards, Guides and Recommendations issued by the World Trade Organization Technical Barriers to Trade (TBT) Committee.*

ASTM F1704, *Standard Test Method for Capture and Containment Performance of Commercial Kitchen Exhaust Ventilation Systems*

Purpose

This test method uses flow visualization (such as schlieren thermal imaging) to determine the threshold of capture and containment (C&C) of a hood/appliance combination under cooking and idle conditions. In laboratory settings, appliances, hoods and replacement air systems can be evaluated to gauge impact on capture and containment performance for a given configuration. Rather than relying on cooking effluent and smoke as a medium alone, this method of test enables the user to gauge performance on the ability to capture the thermal plume emitted from a series of appliances.

1. Scope

1.1 Characterization of capture and containment performance of hood, appliance(s), and replacement air system during cooking and non-cooking conditions (idle):

1.2 Parametric evaluation of operational or design variations in appliances, hoods, or replacement air configurations.

1.3 The test method to determine heat gain to space from commercial kitchen ventilation/appliance systems has been re-designated as Test Method F2474.

1.4 The values stated in inch-pound units are to be regarded as standard. No other units of measurement are included in this standard.

1.5 *This standard does not purport to address all of the safety concerns, if any, associated with its use. It is the responsibility of the user of this standard to establish appropriate safety and health practices and determine the applicability of regulatory limitations prior to use.*

1.6 *This international standard was developed in accordance with internationally recognized principles on standardization established in the Decision on Principles for the Development of International Standards, Guides and Recommendations issued by the World Trade Organization Technical Barriers to Trade (TBT) Committee.*

ASTM F2519, *Standard Test Method for Grease Particle Capture Efficiency of Commercial Kitchen Filters and Extractors*

Purpose

This test method uses a standardized test aerosol ranging in size from 0.3 to 10 microns to characterize capture efficiency of a grease filter as a function of particle size. This method allows more accurate comparison of filter types. Note that grease vapor is not considered in the method of test.

The particulate capture efficiency can be used with known particulate size emission data for a cooking appliance-food product combination to determine the total mass of grease particles captured by the filter, the total mass of grease particles that pass through the filter, and the particle size distribution of the grease particles that pass through the filter.

1. Scope
   - 1.1 This test method can be used to determine the grease particle capture efficiency of components and systems used in commercial kitchens to capture grease effluent prior to entering the exhaust duct. The results can be used to select a filter system best suited to a particular application.
   - 1.2 This test method is applicable to filter components and systems. The performance information is obtained for new or clean filters and does not include the performance of used or loaded filters.
   - 1.3 The filter can be evaluated with respect to the following (where applicable):
     - 1.3.1 Pressure drop as a function of airflow through the filter, and
     - 1.3.2 Particulate capture efficiency by particle size.
   - 1.4 The values stated in inch-pound units are to be regarded as standard. The values given in parentheses are for information only.
   - 1.5 *This test method may involve hazardous materials, operations, and equipment. This standard does not purport to address all of the safety concerns, if any, associated with its use. It is the responsibility of the user of this standard to establish appropriate safety and health practices and determine the applicability of regulatory limitations prior to use.*

ASTM F2976, *Standard Practice for Determining the Field Performance of Commercial Kitchen Demand Control Ventilation Systems*

Purpose

This test method enables evaluation of a demand control ventilation system when installed in new construction or retrofit settings to determine fan, heating and cooling energy savings.

1. Scope
   - 1.1 This practice determines the energy savings potential of Commercial Kitchen Demand Control Ventilation (CKDCV) systems by outlining a procedure to measure system performance.
     - 1.1.1 Fan energy savings potential of a Commercial Kitchen Demand Control Ventilation system will be determined.
     - 1.1.2 Thermal energy savings potential of a Commercial Kitchen Demand Control Ventilation system will be determined.
   - 1.2 This Standard Practice applies to commercial kitchen exhaust and supply demand control ventilation system in the following foodservice establishments: Casino hotel foodservice facilities, commercial cafeterias, full service restaurant, hotel foodservice

facility, quick service restaurant, school cafeteria, supermarket, health care foodservice facility.

1.3 This CKDCV field test protocol does not apply to other demand control ventilation applications such as building heating, ventilation, and air-conditioning (HVAC) applications or laboratory applications.

1.4 Units—The values stated in inch-pound units are to be regarded as standard. The values given in parentheses are mathematical conversions to SI units that are provided for information only and are not considered standard.

1.5 *This standard does not purport to address all of the safety concerns, if any, associated with its use. It is the responsibility of the user of this standard to establish appropriate safety, health, and environmental practices and determine the applicability of regulatory limitations prior to use.*

1.6 *This international standard was developed in accordance with internationally recognized principles on standardization established in the Decision on Principles for the Development of International Standards, Guides and Recommendations issued by the World Trade Organization Technical Barriers to Trade (TBT) Committee.*

ICC Codes and Standards

IgCC/189.1, *International Green Construction Code®*

1. Purpose

   1.1 The purpose of this standard is to provide minimum requirements for the siting, design, construction, and plans for operation of *high-performance green buildings* to

   a. reduce emissions from buildings and building systems, enhance building occupant health and comfort, conserve water resources, protect local biodiversity and ecosystem services, promote sustainable and regenerative materials cycles, enhance building quality, and enhance resilience to natural, technological, and human-caused hazards; and

   b. support the goal of development that meets the needs of the present without compromising the ability of future generations to meet their own needs.

   1.2 This standard is intended to provide the technical basis of mandatory building codes and regulations for *high-performance green buildings* that are broadly adoptable by national and local jurisdictions.

2. Scope

   2.1 This standard contains requirements that

   a. apply to the following *building projects*:
      1. New buildings and their systems
      2. New portions of buildings and their systems
      3. New systems and equipment in existing buildings
      4. Relocated existing buildings and temporary structures where specified in this standard

   b. address *site* sustainability, water use efficiency, energy efficiency, indoor environmental quality (IEQ), materials and resources, and construction and plans for operation.

   2.2 The provisions of this standard do not apply to

   a. single-family houses, multifamily structures of three stories or fewer above grade, manufactured houses (mobile homes), and manufactured houses (modular) and

   b. *building projects* that use none of the following: electricity, fossil fuel, or water.

2.3 The requirements in this standard shall not be used to circumvent any applicable safety, health, or environmental requirements.

International Mechanical Code

[A] 101.3 Purpose

The purpose of this code is to establish minimum requirements to provide a reasonable level of safety, health, property protection and general welfare by regulating and controlling the design, construction, installation, quality of materials, location, operation and maintenance or use of mechanical *equipment* or systems.

[A] 101.2 Scope

This code shall regulate the design, installation, maintenance, *alteration* and inspection of mechanical systems that are permanently installed and utilized to provide control of environmental conditions and related processes within buildings. This code shall also regulate those mechanical systems, system components, *equipment* and *appliances* specifically addressed herein. The installation of fuel gas distribution piping and *equipment*, fuel gas-fired *appliances* and fuel gas-fired *appliance* venting systems shall be regulated by the *International Fuel Gas Code*.

Exception: Detached one- and two-family dwellings and townhouses not more than three stories above grade plane in height with a separate means of egress and their accessory structures not more than three stories above grade plane in height shall comply with this code or the *International Residential Code*.

NFPA Standards

NFPA 13, *Standard for the Installation of Sprinkler Systems*

1.2 Purpose

1.2.1 The purpose of this standard shall be to provide a reasonable degree of protection for life and property from fire through standardization of design, installation, and testing requirements for sprinkler systems, including private fire service mains, based on sound engineering principles, test data, and field experience.

1.1 Scope

1.1.1 This standard shall provide the minimum requirements for the design and installation of automatic fire sprinkler systems and exposure protection sprinkler systems covered within this standard.

1.1.2 This standard shall not provide requirements for the design or installation of water mist fire protection systems.

1.1.2.1 Water mist fire protection systems shall not be considered fire sprinkler systems.

1.1.2.2 The design and installation of water mist fire protection systems shall comply with NFPA 750.

1.1.3 This standard is written with the assumption that the sprinkler system shall be designed to protect against a single fire originating within the building.

NFPA 17A, *Standard for Wet Chemical Extinguishing Systems*

1.2 Purpose. This standard is prepared for the use and guidance of those charged with the purchasing, designing, installation, testing, inspecting, approving, listing, operating, or maintaining of pre-engineered wet chemical fire-extinguishing systems in order that suck equipment will function as intended throughout its life.

1.1 Scope. The provisions of this standard apply to the design, installation, operation, testing, and maintenance of preengineered wet chemical fire extinguishing systems that discharge wet chemical from fixed nozzles and piping by means of expellant gas. It contains only the essential requirements and recommendations needed to make the standard workable in the hands of those skilled in this field.

A.1.1. The wet chemical systems described in this standard are designed to discharge wet chemical from fixed nozzles and piping by means of expellant gas. The intent of the standard is to present the design considerations applicable to these systems.

The wet chemicals produced by various manufacturers usually are not identical in all characteristics, and each manufacturer designs equipment for use with a specific wet chemical. Therefore, system design principles applicable to the products of one manufacturer are not applicable to the products of another manufacturer. As a result, it is not practical to include system design details as part of this standard. However, such system design details are an integral part of the listing of the systems and are included in the manufacturers' design, installation, and maintenance manuals.

NFPA 96, *Standard for Ventilation Control and Fire Protection of Commercial Cooking Operations*

1.2 Purpose. The purpose of this standard shall be to reduce the potential fire hazard of cooking operations, irrespective of the type of cooking equipment used and whether it is used in public or private facilities.

1. Scope.

1.1.1 This standard shall provide the minimum fire safety requirements (preventative and operative) related to the design, installation, operation, inspection, and maintenance of all public and private cooking operations. A.1.1.1 These requirements include, but are not limited to, all manner of cooking equipment, exhaust hoods, grease removal devices, exhaust ductwork, exhaust fans, dampers, fire-extinguishing equipment, and all other auxiliary or ancillary components or systems that are involved in the capture, containment, and control of grease-laden cooking effluent.

1.1.2 This standard shall apply to residential cooking equipment used for commercial cooking operations.

1.1.3 This standard shall not apply to cooking equipment located in a single dwelling unit.

1.1.4 This standard shall not apply to facilities where all of the following are met:

(1) Only residential equipment is being used.

(2) Fire extinguishers are located in all kitchen areas in accordance with NFPA 10, Standard for Portable Fire Extinguishers.

(3) The facility is not an assembly occupancy.

(4) The authority having jurisdiction has approved the installation. A.1.1.4 This judgment should take into account the type of cooking being performed, the items being cooked, and the frequency of cooking operations. Examples of operations that might not require compliance with this standard include the following: (1) Day care centers warming bottles and lunches (2) Therapy cooking facilities in health care occupancies (3) Churches and meeting operations that are not cooking meals that produce grease-laden vapors (4) Employee break rooms where food is warmed.

UL Standards

UL 197, *Commercial Electric Cooking Appliances*

Purpose

This standard establishes safety requirements for commercial electric cooking equipment. Additionally, requirements for appliances with integral systems that limit grease emissions are established which are typically used when for unhooded or ventless applications. Equipment listed with a means to limit grease emissions are classified under the UL product category code KNLZ and do not contain an integral fire extinguishing system.

1. Scope
   - 1.1 These requirements cover commercial electric cooking appliances rated 600 volts or less, intended for indoor use, and intended for use in accordance with the National Electrical Code, NFPA 70.
   - 1.2 These requirements cover coffee makers, conductive cookers, food warmers, fryers, griddles, steam kettles, steam cookers, nut warmers, popcorn machines, ranges, utensil warmers, and other appliances found in commercial kitchens, restaurants, or other business establishments where food is dispensed.
   - 1.3 These requirements do not cover vending machines, cooking appliances intended for household use, commercial cooking appliances rated more than 600 volts, or microwave cooking appliances.
   - 1.4 An appliance designed so that it can be mounted and supported at an outlet box, such as a food warmer, is judged on the basis of compliance with the requirements in this Standard and with the mounting and weight requirements for electric lighting fixtures.
   - 1.5 An appliance that uses heat produced by a means other than electrical (for example, gelled or liquid fuel, coal, gas, or oil) is also investigated with respect to the additional risk of fire.
   - 1.6 An appliance that utilizes heat produced by gelled or liquid fuel is also investigated with the appliance operating with the specific fuel(s) which it intended to be used.
   - 1.7 Commercial cooking equipment employing integral systems for limiting the emission of grease laden air are intended to be installed in accordance with the Standard for Ventilation Control and Fire Protection of Commercial Cooking Operations, NFPA 96, and shall comply with the requirements of this Standard. Additionally, the effluent emitted from the system shall not exceed 5 mg/m^3 as measured in accordance with the U. S. Environmental Protection Agency (EPA) Test Method 202, Determination of Condensable Particulate Emissions From Stationary Sources.
   - 1.8 Commercial cooking equipment employing integral recirculating systems shall comply with the requirements of this Standard and the requirements in the Standard for Recirculating Systems, UL 710B.

UL 300, *Fire Testing of Fire Extinguishing Systems for Protection of Commercial Cooking Equipment*

1. Scope
   - 1.1 These requirements cover the performance during fire tests of pre-engineered fire extinguishing system units intended for the protection of commercial cooking equipment for restaurants, cafeterias and other similar venues. For installation requirements, see the following documents: NFPA 17, Standard for Dry Chemical Extinguishing Systems; NFPA 96, Standard for Ventilation Control and Fire Protection of Commercial Cooking Operations; NFPA 17A, Standard for Wet Chemical

Extinguishing Systems; and the National Building Code of Canada. Note that local authorities having jurisdiction should be consulted prior to installation.

1.2 Pre-engineered fire extinguishing system units, agents, or both shall also comply with the requirements for construction and performance as applicable to specific types, designs, sizes, and arrangements. Such additional requirements include ANSI/CAN/UL/ULC 1254 and are not within the scope of these requirements for fire tests.

1.3 The term "product" as used in this standard refers to all fire extinguishing systems or any part thereof covered by the requirements unless specifically noted otherwise.

UL 705, *Power Ventilators*

Purpose

This standard covers the electrical safety of power ventilations (fans) that are permanently mounted in commercial, industrial and residential applications.

1. Scope

1.1 These requirements cover power ventilators of the roof-and wall-mounted types and duct fans of the straight-through type intended for commercial or industrial use, residential fans intended for heated and conditioned air and for connection to permanently installed wiring systems in accordance with the National Electrical Code, NFPA 70.

1.2 These requirements also cover dryer exhaust duct power ventilators (DEDPV) for single residential dryers.

1.2.1 These requirements cover roof or wall-mounted ventilators and duct fans of the straight-through type for restaurant exhaust appliances.

1.2.2 These requirements cover power ventilators for restaurant exhaust appliances.

UL 710, *Standard for Exhaust Hoods for Commercial Cooking Equipment*

Purpose

This standard establishes product safety requirements for Type I commercial kitchen exhaust hoods when installed over light, medium, heavy and extra heavy-duty equipment. Minimum capture and containment exhaust airflow rates are established in a laboratory setting as well as a means to evaluate an exhaust hood for reduced clearance to combustible materials.

1. Scope

1.1 These requirements cover Type I commercial kitchen exhaust hoods intended for placement over commercial cooking equipment. Exhaust hoods with and without exhaust dampers are covered by these requirements.

1.2 Exhaust hoods with and without exhaust fire actuated fire dampers are covered by these requirements.

1.3 Exhaust hoods are evaluated relative to minimum exhaust air flow required and maximum supply air flow allowed for capture and containment of cooking effluents under laboratory conditions.

1.4 Exhaust hoods may incorporate non-continuous welded joints, seams, and penetrations when evaluated by these requirements.

1.5 Exhaust hoods with fire actuated fire exhaust dampers are intended to have the exhaust fire actuated dampers automatically close to prevent exhaust duct gas temperatures from exceeding 375°F (191°C).

1.6 All exhaust hoods are intended for use with fire extinguishing system units.

1.7 These requirements cover exhaust hoods provided with manually or automatically operated cleaning or washing systems. These requirements do not cover the fire extinguishing aspects of such systems.

1.8 These requirements do not cover evaluation of Ultra Violet (UV) systems for use in commercial kitchen exhaust systems. The Outline for Ultraviolet Radiation Systems for Use in the Ventilation Control of Commercial Cooking Operations, UL 710C, covers these products.

1.9 These requirements do not cover evaluation of Electrostatic Precipitators (ESP's) for use in commercial kitchen ventilation. Electrostatic Precipitators (ESP's) are covered under the Standard for Electrostatic Air Cleaners, UL 867.

1.10 These requirements do not cover evaluation of commercial electric cooking appliances provided with integral recirculating systems (previously referred to as ductless hoods) and nonintegral recirculating systems, both of which are intended for installation in commercial establishments for the preparation of food. The Standard for Recirculating Systems, UL 710B, covers these products.

1.11 Exhaust hoods covered by these requirements are intended for installation in accordance with the following:

a) The Standard for Ventilation Control and Fire Protection of Commercial Cooking Operations, NFPA 96;
b) The National Electrical Code, NFPA 70; and/or
c) Other codes such as the International Mechanical Code (IMC) and the Uniform Mechanical Code (UMC).

1.12 These requirements cover products rated 600 volts or less.

1.13 These requirements do not cover evaluation of the exhaust hoods with respect to their grease extraction efficiency.

Note: Capture efficiency of a kitchen hood filter can be measured using ASTM F2519 "Standard Test Method for Grease Particle Capture Efficiency of Commercial Kitchen Filters and Extractors".

UL 710B, *Recirculating Systems*

Purpose

This standard establishes product safety requirements for recirculating or ventless hood systems covering electrical cooking appliances that include a means to interlock operation of the appliances as well as an integral fire suppression system. Clearance to combustible materials for the systems are established as part of the testing process.

1. Scope

1.1 These requirements cover the following recirculating systems rated 600 volts or less and intended for indoor use:

a) Non-integral recirculating systems for installation in commercial establishments for the preparation of food;
b) Electric commercial cooking appliances provided with integral recirculating systems intended for installation in commercial establishments for the preparation of food; and
c) Vending machines provided with integral recirculating systems.

1.2 These devices incorporate an air filtering system enclosed in a hooded or otherwise contained area intended to capture air from the cooking process area. The hood assembly typically includes a fan, collection hood, or equivalent design and an air filtering system (consisting of a grease filter with or without other filtering means). In addition, except as noted in the Exceptions to 37.1 and 38.1, the hood assembly includes a fire actuated damper, and a fire extinguishing system.

1.3 Recirculating systems covered by these requirements are intended for installation in accordance with:

a) The Standard for Ventilation Control and Fire Protection of Commercial Cooking Operations, NFPA 96;
b) The National Electrical Code, ANSI/NFPA 70; and
c) Other codes such as the International Mechanical Code (IMC), and the Uniform Mechanical Code (UMC).

1.4 Integral recirculating systems shall comply with the requirements for the cooking appliance as noted below, except as modified or superseded by the requirements in this Standard.

a) Commercial electric cooking appliances shall comply with the requirements in the Standard for Commercial Electric Cooking Appliances, UL 197.
b) Vending machines shall comply with the requirements in the Standard for Vending Machines, UL 751.

UL 710C, *Outline of Investigation for Ultraviolet Radiation Systems for Use in the Ventilation Control of Commercial Cooking Operations*

Purpose

This standard establishes product safety requirements for ultraviolet radiation lamp systems when installed in commercial exhaust hoods. These systems are required to incorporate interlocks to ensure operators are not exposed to harmful levels of ultraviolet irradiance.

1. Scope

1.1 These requirements cover ultraviolet (UV) radiation lamp systems used for the reduction of grease laden vapors from commercial cooking equipment when installed within Listed Exhaust Hoods for Commercial Cooking Equipment.

1.2 These requirements are not intended to address the effectiveness of the UV radiation to reduce grease-laden vapors from cooking operations.

1.3 This Outline applies to UV lamp assemblies for use in locations that are intended for installation on branch circuits of 600 V nominal or less between conductors in accordance with the National Electrical Code (NEC), ANSI/NFPA 70.

UL 762, *Outline of Investigation for Power Roof Ventilators for Restaurant Exhaust Appliances*

Purpose

This standard establishes product safety requirements for roof or wall mounted exhaust fans with considerations for use in the commercial kitchen environment where elevated temperatures and grease are present in the airstream.

1. Scope

1.1 These requirements cover roof or wall-mounted ventilators and duct fans of the straight-through type for restaurant exhaust appliances.

1.2 Power ventilators for restaurant exhaust appliances covered by these requirements are intended for installation in accordance with the Standard of the National Fire Protection Association for the Installation of Equipment for the Removal of Smoke and Grease-Laden Vapors from Commercial Cooking Equipment, NFPA 96.

1.3 Power ventilators for restaurant appliances shall comply with the Standard for Power Ventilators, UL 705, modified in that wiring system shall not be located in the path of travel of exhaust products.

UL 1046, *Grease Filters for Exhaust Ducts*

Purpose

This standard establishes product safety requirements for grease extractors installed in Type I commercial kitchen exhaust hoods. Filters are evaluated in regard to ability to limit flame propagation downstream into a duct system and draining grease in such a manner that it does not drip onto cooking equipment surfaces. Note that a Type I commercial kitchen exhaust hood may also be fitted with an integral grease extractor that is evaluated as part of the UL710 evaluation.

1. Scope

1.1 Products covered by these requirements are single-stage and multi-stage grease filters. These grease filters are intended for installation and use in accordance with the Standard for Ventilation Control and Fire Protection of Commercial Cooking Operations, NFPA No. 96 and the International Mechanical Code (IMC).

1.2 Grease extraction devices integral to the exhaust hoods that use them and other filter devices designed to be used in only the specific manufacturer's hoods are not covered by this standard but are tested under the Standard for Exhaust Hoods for Commercial Cooking Equipment, UL 710, or the Standard for Recirculating Systems, UL 710B, or the Outline for Ultraviolet Radiation Systems for Use in the Ventilation Control of Commercial Cooking Operations, Subject 710C.

1.3 The test methods described in this standard pertain to filters used in exhaust hoods for commercial cooking operations. The primary function of these devices and the capability of devices tested in this standard are to prevent flames from penetrating into the exhaust system beyond the hood collar. The terms grease and oil are used interchangeably in this standard.

1.4 Filters are investigated to determine their abilities to:

a) Limit the projection of flames downstream when subjected to flames on the upstream face, after having been loaded with grease in a manner representative of cooking that produces a grease-rich exhaust,

b) Drain any collected grease in such a manner that it does not fall back onto the cooking surface.

1.5 The standard does not measure capture efficiency or physical robustness of the grease filter device or other factors that do not directly affect user or fire safety.

Note: Capture efficiency of a kitchen hood filter can be measured using ASTM F2519 "Standard Test Method for Grease Particle Capture Efficiency of Commercial Kitchen Filters and Extractors".

UL 1978, *Grease Ducts*

Purpose

This standard is used to ensure the duct construction of factory-built grease ducts and grease duct assemblies are safe and to determine the minimum reduced clearance to combustibles required with these ducts and assemblies.

1. Scope
   - 1.1 These requirements cover factory-built grease ducts, and grease duct assemblies that are intended to be installed at reduced clearances where 18 inch (457 mm) clearance is specified in the Standard for Ventilation Control and Fire Protection of Commercial Cooking Operations, NFPA 96, and the International Mechanical Code.
   - 1.2 These requirements also cover modular grease duct assemblies, unwelded connections between adjoining duct parts, fittings, access doors, and the like intended for use with grease ducts installed in accordance with NFPA 96 and the International Mechanical Code.
   - 1.3 These requirements do not address NFPA 96 and the International Mechanical Code requirements for grease duct enclosures. Grease ducts covered only by these requirements are intended to be installed in a fire resistive enclosure when required by NFPA 96 or the IMC.
   - 1.4 Requirements used to evaluate the effectiveness of the combination of a grease duct and an enclosure as a fire rated enclosure system and through penetration firestop system, as well as the enclosure's effect on the grease duct are contained in Standard for Tests of Fire Resistive Grease Duct Enclosure Assemblies, UL 2221, and ASTM E 2336, Standard Test Methods for Fire Resistive Grease Duct Enclosure Systems.

UL 2221, *Standard for Tests of Fire Resistive Grease Duct Enclosure Assemblies*

Purpose

The purpose of this standard is to determine the fire resistance of assemblies which cover or enclose the grease duct. Usually these are integral enclosures around factory-built listed grease duct.

1. Scope
   - 1.1 These tests are intended to determine the fire resistance of grease duct enclosure assemblies. These requirements limit the combustibility, the surface flammability, and the smoke generation potential of the coverings used to enclose the grease duct. In addition, these requirements evaluate the effectiveness of the combination of the grease duct and the enclosure as a fire rated enclosure system and through penetration firestop system, as well as the enclosure's effect on the grease duct.
   - 1.2 The fire endurance ratings for grease duct enclosure assemblies are intended to register performance during the period of fire exposure and are not intended to be interpreted as having determined the acceptability of the grease duct assembly for use before or after fire exposure. The intent of these methods is to develop data to assist others in determining the suitability of the grease duct assembly where fire resistance is required.
   - 1.3 These requirements are intended to evaluate the ability of grease duct enclosure assemblies specified in 1.1 to contain an internal fire or repel an external fire during predetermined test exposures. The test evaluates the grease duct assembly's resistance to fire, heat, and to a hose stream.
   - 1.4 Requirements covering the construction and performance requirements of proprietary grease ducts are contained in the Standard for Grease Ducts, UL 1978. All ducts, both proprietary and non-proprietary, shall comply with installation requirements of the Standard for Ventilation Control and Fire Protection of Commercial Cooking Operations, NFPA 96, and the International Mechanical Code.

1.5 Under these requirements a grease duct enclosure assembly is subjected to standard internal and external fire exposures controlled to achieve specified temperatures throughout a specified time period.

1.6 During the internal fire exposure, one of two installation condition methods shall be selected to measure temperatures near or on the surface of the grease duct enclosure material away from the grease duct. One method of temperature measurement is intended for all installations except for installations within non-ventilated combustible enclosures. The second method of temperature measurement is intended for installations within non-ventilated combustible enclosures.

1.7 These exposures by themselves are not intended to be representative of all fire conditions; conditions vary with changes in the amount, nature, and distribution of fire loading, ventilation, compartment size and configuration, and heat sink characteristics of the compartment.

1.8 These requirements cover the measurement of the transmission of heat through the grease duct assembly and the passage of heat and gases hot enough to ignite cotton waste.

1.9 These requirements provide a relative measure of fire performance under these specified fire exposure conditions. Any variation from the construction or conditions that are tested, such as method of assembly and materials, is not within the scope of this test method.

1.10 The results of these tests represent one factor in assessing fire performance of grease duct assemblies. These requirements prescribe standard fire exposures for assessing the performance of grease duct assemblies. Application of these test results to predict the performance of actual building construction requires careful evaluation of test data.

UL 8782, *Outline of Investigation for Pollution Control Units for Commercial Cooking Operations*

Purpose

To establish the minimum safety requirements for a pollution control unit installed to clean the exhaust air stream or perform heat recovery from the duct system of a commercial kitchen.

1. Scope

1.1 These requirements cover commercial pollution control unit assemblies that are intended to be installed in line with commercial grease ducts used to exhaust heat, smoke, and grease-laden vapors from commercial cooking operations to the outdoors as specified in the Standard for Ventilation Control and Fire Protection of Commercial Cooking Operations, NFPA 96, the International Mechanical Code (IMC), and the Uniform Mechanical Code (UMC).

1.2 These requirements include modular pollution control assemblies, unwelded connections between adjoining parts, fittings, access doors, and the like.

1.3 These requirements do not address NFPA 96 and the International Mechanical Code requirements for pollution control unit fire resistive enclosures. Pollution control units covered only by these requirements are intended to be installed in a fire resistive enclosure when required by NFPA 96 or the IMC.

1.4 Requirements used to evaluate the effectiveness of the combination of a pollution control unit and an enclosure as a fire rated enclosure system and through penetration firestop system, as well as the enclosure's effect on the pollution control unit are contained in Standard for Tests of Fire Resistive Grease Duct Enclosure Assemblies, UL 2221, and ASTM E 2336, Standard Test Methods for Fire Resistive Grease Duct Enclosure Systems.

1.5 This Outline does not measure the efficacy of the pollution control unit to remove grease-laden effluent from the air stream, nor other factors that do not directly affect users or electrical and fire safety.

Codes and Standards Citations

ASHRAE. 2019. *ASHRAE handbook—HVAC applications*. Peachtree Corners, GA: ASHRAE.

ASHRAE. 2021. *ASHRAE handbook—Fundamentals*. Peachtree Corners, GA: ASHRAE.

ASHRAE. 2022. ANSI/ASHRAE Standard 62.1-2022, *Ventilation for acceptable indoor air quality*. Peachtree Corners, GA: ASHRAE.

ASHRAE. 2022. ANSI/ASHRAE/IES Standard 90.1-2022, *Energy standard for sites and buildings except low-rise residential buildings*. Peachtree Corners, GA: ASHRAE.

ASHRAE. 2022. ANSI/ASHRAE Standard 154-2022, *Ventilation for commercial cooking operations*. Peachtree Corners, GA: ASHRAE.

ASTM. 2017. ASTM F2975-12(2017), *Standard test method for measuring the field performance of commercial kitchen ventilation systems.* West Conshohocken, PA: ASTM International.

ASTM. 2019. ASTM F2976-13(2019), *Standard practice for determining the field performance of commercial kitchen demand control ventilation systems*. West Conshohocken, PA: ASTM International.

ASTM. 2020. ASTM E2336-20, *Standard test method for fire resistive grease duct enclosure systems*. West Conshohocken, PA: ASTM International.

ASTM. 2020. ASTM F2519-05(2020), *Standard test method for grease particle capture efficiency of commercial kitchen filters and extractors*. West Conshohocken, PA: ASTM International.

ASTM. 2022. ASTM F1704-12(2022), *Standard test method for capture and containment performance of commercial kitchen exhaust ventilation systems*. West Conshohocken, PA: ASTM International.

ICC. 2021. IgCC/189.1. *International green construction code®* powered by ANSI/ASHRAE/ICC/USGBC/IES Standard 189.1. Washington, DC: International Code Council.

ICC 2021. *International mechanical code*. Washington, DC: International Code Council.

IAPMO. 2021. *Uniform mechanical code*. Ontario, Canada: International Association of Plumbing and Mechanical Officials.

NFPA. 2021. NFPA 17A, *Standard for wet chemical extinguishing systems*. Quincy, MA: National Fire Protection Association.

NFPA. 2021. NFPA 96-2017, *Standard for ventilation control and fire protection of commercial cooking operations*. Quincy, MA: National Fire Protection Association.

NFPA. 2022. NFPA 13, *Standard for the installation of sprinkler systems*. Quincy MA: National Fire Protection Association.

UL. 2006. UL 710C, *UL LLC outline of investigation for ultraviolet radiation systems for use in the ventilation control of commercial cooking operations*. Northbrook, IL: UL Solutions.

UL. 2010. UL 197, *Standard for commercial electric cooking appliances*. Northbrook, IL: UL Solutions.

UL. 2010. UL 1046, *Grease filters for exhaust ducts*. Northbrook, IL: UL Solutions.

UL. 2010. UL 1978, *Grease ducts*. Northbrook, IL: UL Solutions.

UL. 2010. UL 2221, *Standard for tests of fire resistive grease duct enclosure assemblies*. Northbrook, IL: UL Solutions.

UL. 2011. UL 710B, *Recirculating systems*. Northbrook, IL: UL Solutions.

UL. 2012. UL 710, *Exhaust hoods for commercial cooking equipment*. Northbrook, IL: UL Solutions.

UL. 2013. UL 762, *UL LLC outline of investigation for power roof ventilators for restaurant exhaust appliances*. Northbrook, IL: UL Solutions.

UL. 2017. UL 705, *Power ventilators*. Northbrook, IL: UL Solutions.

UL. 2017. UL 8782, *UL LLC outline of investigation for pollution control units for commercial cooking operations*. Northbrook, IL: UL Solutions.

UL. 2019. UL 300, *Standard for fire testing of fire extinguishing systems for protection of commercial cooking equipment*. Northbrook, IL: UL Solutions.

B

Research Conducted by ASHRAE TC 5.10

Research Overview

ASHRAE Technical Committee (TC) 5.10, Kitchen Ventilation, has been active for many years in research related to kitchen ventilation. This research has tended to focus on answering questions related to field-related issues such as measuring exhaust airflow rates for hood and replacement air systems (RP-623 and RP-1376), how much grease is produced by hooded cooking appliances (RP-745 and RP-1375), and evaluating the grease and heat gain from unhooded countertop cooking appliances (RP-1631). Some of the research has focused on the design aspects of kitchen ventilation systems, from optimizing exhaust hood performance (RP-1202 and RP-1480) to evaluating the grease removal efficiency of filtering devices (RP-851 and RP-1151) and reducing the velocity of airflow in the exhaust ductwork (RP-1033). Other projects, such as RP-1362, have evaluated the relationships between appliances and ventilation systems and the HVAC system in the space.

RP-1469 is a comprehensive study that was conducted using both field measurements and field surveys regarding thermal comfort in commercial kitchens. RP-1614 was completed with the goal of developing a method of test to determine the effectiveness of UVC systems installed in commercial kitchen ventilation (CKV) systems.

RP-1778 studied the heat and moisture loads in commercial kitchen dishrooms. This research provides invaluable information to engineers, HVAC designers, and consultants on the actual loads to the dishroom or kitchen space.

ASHRAE members can download ASHRAE Research Project (RP) research reports from the ASHRAE Technology Portal at www.ashrae.org/technologyportal.

Benefits to the HVAC Industry

Many research projects sponsored by TC 5.10 have impacted energy use and sustainability in the food service industry. The research from RP-1033 provided data that show grease deposition on the duct walls actually decreased when the duct velocity was lowered from 1500 to 500 fpm (7.6 to 2.5 m/s). This data allowed both NFPA 96, *Standard for Ventilation Control and Fire Protection of Commercial Cooking Operations* (2021b), and *International Mechanical Code* (IMC; ICC 2021d) to allow lower duct velocities. These changes allow demand-controlled ventilation systems (where the airflow is lowered during non-cooking periods of the day) to be used at a national level to achieve significant energy savings.

The two TC 5.10 research projects related to hood performance, RP-1202 and RP-1480, not only evaluated how wall canopy and island hoods perform with various appliances but also methods of reducing exhaust airflows required for hoods to capture cooking effluent more efficiently. These methods include such options as optimizing the appliances' positions underneath the hoods, installing side panels, and designing hoods to use larger overhangs if possible. Reducing the exhaust air generally reduces the amount of conditioned air needing to be brought back into the space to replace the air that is exhausted, thus leading to large energy savings in commercial kitchens. RP-1362 measured the heat gain from appliances underneath hoods. These data can be used to more accurately size the HVAC equipment needed to condition the kitchen space.

Earlier projects related to grease emissions (RP-851, RP-745, and RP-1151) were used to help develop ASTM F2519, *Standard Test Method for Grease Particle Capture Efficiency of Commercial Kitchen Filters and Extractors* (2020b). The data from these research projects, along with ASTM F2519 and the data from RP-1375, have revolutionized the kitchen ventilation industry with regard to how mechanical filters actually perform in the field. ASTM F2519 provides a framework for making more efficient filters that help reduce the amount of grease buildup in ductwork, on exhaust fans, and on the roofs of buildings.

RP-1631 evaluated the appliance emissions and heat gain to the space from countertop commercial cooking appliances to help determine whether these cooking processes require ventilation hoods or can be vented to the space.

RP-623 and RP-1376 both tackle how to accurately measure the exhaust and replacement air in food service establishments. By being able to more accurately measure the airflows, restaurants can be properly balanced to the design conditions so that excess energy that is not needed is not being consumed.

For a comprehensive list of TC 5.10 research projects, see Table B.1. Complete citations for all the ASHRAE research reports mentioned in this chapter are included in the References of this book.

Table B.1 ASHRAE TC 5.10 Research Projects through 2023

Years	ASHRAE Research Project	Title	Authors
1993–1994	RP-623	Testing Grease Hoods	Eliott B. Gordon, Fuoad Parvin
1996–1997	RP-851	Grease Extraction and Removal from Exhaust Air Streams of Cooking Processes	Dereck Schrock, Kevin Knappmiller
1998–1999	RP-745	Identification and Characterization of Cooking Effluents as Related to Optimum Design of Kitchen Ventilation Systems	Thomas Kuehn, William D. Gerstler, David Y.H. Pui, James W. Ramsey, Michael P. Doerr
2000–2001	RP-1033	Effects of Air Velocity on Grease Deposition in Exhaust Ductwork	Thomas H. Kuehn, William Gerstler, Humberto Orriz, Aaron Sandquist, Handi Tjandra, Jitesh J. Vidhani, James W. Ramsey, David Y.H. Pui
2001–2003	RP-1151	Development of a Draft Method of Test for Determining Grease Removal Efficiencies	William A. Welch, Thomas H. Kuehn
2003–2005	RP-1202	Effect of Appliance Diversity and Position on Commercial Kitchen Hood Performance	Richard Swierczyna, Paul Sobiski, Donald Fisher
2007–2008	RP-1375	Characterization of Effluents from Additional Cooking Appliances	Thomas H. Kuehn, Bernard A. Olson, James W. Ramsey, Joshua M. Rocklage
2008–2009	RP-1362	Revised Heat Gain and Capture and Containment Exhaust Rates from Typical Commercial Cooking Appliances	Donald Fisher, Richard Swierczyna, Paul Sobiski
2008–2009	RP-1480	Island Hood Energy Consumption and Energy Reduction Strategies	Richard Swierczyna, Paul Sobiski, Donald Fisher
2008–2010	RP-1376	Method of Test to Evaluate Field Performance of Commercial Kitchen Ventilation Systems	Thomas H. Kuehn, Bernard A. Olson, Daniel Bissell, Kevin Campbell, Andrew Hawkinson
2010–2012	RP-1469	Comfort in Commercial Kitchens	John Stoops, Amber Watkins, Ed Smyth, Mark Adams, A. Simone, B.W. Olesen
2013–2015	RP-1631	Countertop Commercial Appliance Emissions	Jianshun Zheng, Kwanghoon Han, Bing Guo, Yan Zhang, Changfu Wu
2017–2019	RP-1614	Developing a Test Method to Determine the Effectiveness of UVC Systems on Commercial Cooking Effluent	Jianshun "Jensen" Zheng, Meng Kong, Kwan Hoon Han, Bing Guo, Zhenlei Liu
2018–2020	RP-1778	Heat and Moisture Load from Commercial Dishroom Appliances and Equipment	Denis Livchak, Rich Swierczyna

References

AABC. 2016. *National standards for total system balance*. Washington, D.C.: Associated Air Balance Council.

ACGIH. 2023. TLV/BEI guidelines. Cincinnati, OH: American Conference of Governmental Industrial Hygienists. www.acgih.org/science/tlv-bei-guidelines/.

ADA. 2023. Americans with Disabilities Act of 1990, As Amended. Washington, D.C.: Americans with Disabilities Act. www.ada.gov/law-and-regs/ada/.

ASHRAE. 2017. ASHRAE Standard 52.2-2017, *Method of testing general ventilation air-cleaning devices for removal efficiency by particle size*. Peachtree Corners, GA: ASHRAE.

ASHRAE. 2018. ANSI/ASHRAE/IES Standard 202-2018, *Commissioning process for buildings and systems*. Peachtree Corners, GA: ASHRAE.

ASHRAE. 2019. *ASHRAE handbook—HVAC applications*. Peachtree Corners, GA: ASHRAE.

ASHRAE. 2020. ANSI/ASHRAE Standard 55-2020, *Thermal environmental conditions for human occupancy*. Peachtree Corners, GA: ASHRAE.

ASHRAE. 2021a. ANSI/ASHRAE Standard 169-2021, *Climatic data for building design standards*. Peachtree Corners, GA: ASHRAE.

ASHRAE. 2021b. *ASHRAE handbook—Fundamentals*. Peachtree Corners, GA: ASHRAE.

ASHRAE. 2022a. *ASHRAE position document on infectious aerosols*. Peachtree Corners, GA: ASHRAE. www.ashrae.org/file%20library/about/position%20documents/pd_infectiousaerosols_2020.pdf.

ASHRAE. 2022b. ANSI/ASHRAE/IES Standard 90.1-2022, *Energy standard for sites and buildings except low-rise residential buildings*. Peachtree Corners, GA: ASHRAE.

ASHRAE. 2022c. ANSI/ASHRAE Standard 62.1-2022, *Ventilation for acceptable indoor air quality*. Peachtree Corners, GA: ASHRAE.

ASHRAE. 2022d. ANSI/ASHRAE Standard 154-2022, *Ventilation for commercial cooking operations*. Peachtree Corners, GA: ASHRAE.

ASTM. 2017. ASTM E814-13a, *Standard test method for fire tests of penetration firestop systems*. West Conshohocken, PA: ASTM International.

ASTM. 2017. ASTM F2975-12(2017), *Standard test method for measuring the field performance of commercial kitchen ventilation systems*. West Conshohocken, PA: ASTM International.

ASTM. 2019. ASTM F2976-13(2019), *Standard practice for determining the field performance of commercial kitchen demand control ventilation systems*. West Conshohocken, PA: ASTM International.

ASTM. 2020a. ASTM E2336-20, *Standard test method for fire resistive grease duct enclosure systems*. West Conshohocken, PA: ASTM International.

ASTM. 2020b. ASTM F2519-05(2020), *Standard test method for grease particle capture efficiency of commercial kitchen filters and extractors*. West Conshohocken, PA: ASTM International.

ASTM. 2022. ASTM F1704-12(2022), *Standard test method for capture and containment performance of commercial kitchen exhaust ventilation systems*. West Conshohocken, PA: ASTM International.

Brohard, G., D.R. Fisher, V. Smith, R. Swierczyna, and P. Sobiski. 2003. *Makeup air effects on kitchen exhaust hood performance*. Sacramento, CA: California Energy Commission.

CDC. 2018. Notes from the field: Occupational carbon monoxide exposure in an industrial kitchen facility—Wisconsin, 2017. *Morbidity and Mortality Weekly Report* 67(28):786. www.cdc.gov/mmwr/volumes/67/wr/mm6728a5.htm.

CEC. 2002. Design Guide 3, *Optimizing makeup air: Improving commercial kitchen ventilation (CKV) system performance*. Sacramento, CA: California Energy Commission. https://caenergywise.com/design-guides/CKV-Design-Guide-3_Optimizing_Makeup_Air.pdf.

CEC. 2019. Chapter 13, Acceptance test requirements. *2019 nonresidential compliance manual*. Sacramento, CA: California Energy Commission.

CEC. 2022. *2022 California building standards code* (Cal. code regs., Title 24). Sacramento, CA: California Energy Commission.

Clark, J.C. 2006. If you can't stand the heat, it's time to fix the kitchen. *ASHRAE Journal* 48(4):60–66.

CSA Group. 2018. CAN/CSA 22.2 No. 113-18, *Fans and ventilators*. Toronto, Ontario, Canada: CSA Group.

DOE. 2015. Guidance on demand-controlled kitchen ventilation. Washington, D.C.: U.S. Department of Energy. https://betterbuildingssolutioncenter.energy.gov/sites/default/files/attachments/Guidance-on-Demand-Controlled Kitchen-Ventilation.pdf.

EPA. 2013. Moisture control guidance for building design, construction, and maintenance. Washington, D.C.: U.S. Environmental Protection Agency. www.epa.gov/sites/production/files/2014-08/documents/moisture-control.pdf.

EPA. 2016. *EPA Method 202 Best Practices Handbook*. Washington, D.C.: U.S. Environmental Protection Agency.

EPA. 2022. Chapter 1, Introduction. *Mold remediation in schools and buildings guide*. Washington, D.C.: U.S. Environmental Protection Agency. www.epa.gov/mold/mold-remediation-schools-and-commercial-buildings-guide.

European Standards. 2019. CSN EN 1822-1, *High efficiency air filters (EPA, HEPA and ULPA)—Part 1: Classification, performance testing, marking*. Czech Republic: European Standards.

Fisher, D., R. Swierczyna, and P. Sobiski. 2008. *Revised heat gain and capture and containment exhaust rates from typical commercial cooking appliances*. ASHRAE Research Project RP-1362 Final Report. Peachtree Corners, GA: ASHRAE.

Fisher, D., R. Swierczyna, and A. Karas. 2013. Future of DCV for commercial kitchens. *ASHRAE Journal* 55(2):48–54.

Fuller, S., and S. Petersen. 1996. *Life-cycle costing manual for the Federal Energy Management Program*. NIST handbook 135, 1995 ed. Gaithersburg, MD: National Institute of Standards and Technology.

Gerstler, W., T. Kuehn, D. Pui, J. Ramsey, M. Doerr. 1998. *Identification and characterization of cooking effluents as related to optimum design of kitchen ventilation systems*. ASHRAE Research Project RP-745 Final Report. Peachtree Corners, GA: ASHRAE.

GTM. 2015. Carbon monoxide detectors required in New York businesses. Clifton Park, NY: GTM Payroll Services Inc. https://gtm.com/business/carbon-monoxide-detectors-required-in-new-york/#:~:text=Effective%20June%2027%2C%202015%2C%20all,family%20homes%2C%20condominiums%20and%20cooperatives.

IAPMO. 2021. *Uniform mechanical code*. Ontario, Canada: International Association of Plumbing and Mechanical Officials.

ICC. 2021a. IgCC/189.1, *International green construction code®* powered by ANSI/ASHRAE/ICC/USGBC/IES Standard 189.1. Washington, D.C.: International Code Council.

ICC. 2021b. *International building code*. Washington, D.C.: International Code Council.

ICC. 2021c. *International energy conservation code*. Washington, D.C.: International Code Council.

ICC. 2021d. *International mechanical code*. Washington, D.C.: International Code Council.

ISO. 2016. ISO 16890-1, *Air filters for general ventilation—Part 1: Technical specifications, requirements and classification system based upon particulate matter efficiency (ePM)*. Geneva, Switzerland: International Organization for Standardization.

ISO. 2017. ISO 29463-1, *High efficiency filters and filter media for removing particles from air—Part 1: Classification, performance, testing and marking*. Geneva, Switzerland: International Organization for Standardization.

IWBI. n.d. WELL certification. New York, NY: International WELL Building Institute. www.wellcertified.com/ certification/v2.

Johnson-Tate, D.R. 2018. Effective strategies used by restaurant managers to reduce employee absenteeism. Dissertation, Walden University.

Kleeman, M.J., J.J. Schauer, and G.R. Cass. 1999. Size and composition distribution of fine particulate matter emitted from wood burning, meat charbroiling, and cigarettes. *Environmental Science and Technology* 33(20):3516–23.

Kuehn, T.H., W.D. Gerstler, H. Orriz, A. Sandquist, H. Tjandra, J.J. Vidhani, D.Y.H. Pui, and J.W. Ramsey. 2000. *Effects of air velocity on grease deposition in ductwork*. ASHRAE Research Project RP-1033 Final Report. Peachtree Corners, GA: ASHRAE.

Kuehn, T.H., B.A. Olson, J.W. Ramsey, and J.M. Rocklage. 2008. *Characterization of effluents from additional cooking appliances*. ASHRAE Research Project RP-1375 Final Report. Peachtree Corners, GA: ASHRAE.

Kuehn, T.H., B.A. Olson, D. Bissell, K. Campbell, and A. Hawkinson. 2010. *Method of test to evaluate field performance of commercial kitchen ventilation systems*. ASHRAE Research Project RP-1376 Final Report. Peachtree Corners, GA: ASHRAE.

Lawrence Berkeley National Laboratory. 2007. Berkeley Lab, EPA studies confirm large public health and economic impact of dampness and mold. Berkley, CA: Lawrence Berkeley National Laboratory. www2.lbl.gov/Science-Articles/Archive/EETD-mold-risk.html.

Livchak, A., D. Schrock, and S. Zeqiang. 2005. The effect of supply air systems on kitchen thermal environment. *ASHRAE Transactions* 111(1):748–54.

Livchak, A., D. Schrock, M. Lehtimaki, and A. Talpale. 2003. The facts about mechanical grease filters. *Clearing the air on kitchen ventilation: A supplement to ASHRAE Journal*. Peachtree Corners, GA: ASHRAE.

Livchak, D., and R. Swierczyna. 2020. *Heat and moisture load from commercial dishroom appliances and equipment*. ASHRAE Research Project RP-1778 Final Report. Peachtree Corners, GA: ASHRAE.

NEBB. 2019. *Procedural standard for testing, adjusting and balancing of environmental systems*. Gaithersburg, MD: National Environmental Balancing Bureau.

NFPA. 2021a. NFPA 17A, *Standard for wet chemical extinguishing systems*. Quincy, MA: National Fire Protection Association.

NFPA. 2021b. NFPA 96, *Standard for ventilation control and fire protection of commercial cooking operations*. Quincy, MA: National Fire Protection Association.

NFPA. 2022. NFPA 13, *Standard for the installation of sprinkler systems*. Quincy MA: NFPA.

NSF. 2021. NSF/ANSI 2-2021, *Food equipment*. Ann Arbor, MI: National Science Foundation.

OSHA. 2023. Section III: Chapter 3, Ventilation investigation. *OSHA technical manual (OTM)*. Washington, D.C.: Occupational Safety & Health Administration. www.osha.gov/otm/section-3-health-hazards/chapter-3.

SAFE. 2018. Understanding the health problems associated with black mold. Vallejo, CA: Safe Air Fast Environmental. www.safeairfast.com/blog/2014/08/understanding-the-health-problems-associated-with-black-mold/.

Schrock, D., J. Sandusky, and A. Livchak. 2012. Demand-controlled ventilation for commercial kitchens. *ASHRAE Journal* 54(12):36–48.

Simone, A., B.W. Olesen, J.L. Stoops, and A.W. Watkins. 2013. Thermal comfort in commercial kitchens (RP-1469): Procedure and physical measurements (Part 1). *HVAC&R Research* 19(8):1001–15.

Singh, A., R. Kamal, M.K.R. Mudiam, M.K. Gupta, G.N.V. Satyanarayana, V. Bihari, N. Shukla, A.H. Khan, and C.N. Kesavachandran. 2016. Heat and PAHs emissions in indoor kitchen air and its impact on kidney dysfunctions among kitchen workers in Lucknow, North India. *PLoS One* 11(2). www.ncbi.nlm.nih.gov/pmc/articles/PMC4752274/.

SMACNA. 2020. *HVAC duct construction standards—Metal and flexible*, 4th ed. Chantilly, VA: Sheet Metal and Air Conditioning Contractors' National Association.

Stoops, J., A. Watkins, E. Smyth, M. Adams, A. Simone, and B.W. Olesen. 2013. *Comfort in commercial kitchens*. ASHRAE Research Project RP-1469 Final Report. Peachtree Corners, GA: ASHRAE.

Swierczyna, R., P. Sobiski, and D.Fisher. 2005. *Effect of appliance diversity and position on commercial kitchen hood performance*. ASHRAE Research Project RP-1202 Final Report. Peachtree Corners, GA: ASHRAE.

Swierczyna, R., P. Sobiski, and D. Fisher. 2010. *Island hood energy consumption and energy consumption strategies*. ASHRAE Research Project RP-1480 Final Report. Peachtree Corners, GA: ASHRAE.

Tabunshchikov, I.A., A.N. Kolubkov, M.M. Brodach, and I.S. Avakian. 2020. Calculating airflow rates, cooling loads in commercial kitchens. *ASHRAE Journal* 62(9):48–50.

Trane. 2021. TRACE® 3D Plus. www.trane.com/commercial/north-america/us/en/products-systems/design-and-analysis-tools/trane-design-tools/trace-3d-plus.html.

UL. 2006. UL 710C, *UL LLC outline of investigation for ultraviolet radiation systems for use in the ventilation control of commercial cooking operations*. Northbrook, IL: UL Solutions.

UL. 2010a. UL 103, *Factory-built chimneys for residential type and building heating appliances*. Bensenville, IL: UL Solutions.

UL. 2010b. UL 197, *Standard for commercial electric cooking appliances*. Northbrook, IL: UL Solutions.

UL. 2010c. UL 1046, *Grease filters for exhaust ducts*. Northbrook, IL: UL Solutions.

UL. 2010d. UL 1978, *Grease ducts*. Northbrook, IL: UL Solutions.

UL. 2010e. UL 2221, *Standard for tests of fire resistive grease duct enclosure assemblies*. Northbrook, IL: UL Solutions.

UL. 2011a. UL 710B, *Recirculating systems*. Northbrook, IL: UL Solutions.

UL. 2011b. UL 867, *Electrostatic air cleaners*. Bensenville, IL: UL Solutions.

UL. 2012. UL 710, *Exhaust hoods for commercial cooking equipment*. Bensenville, IL: UL Solutions.

UL. 2013. UL 762, *UL LLC outline of investigation for power roof ventilators for restaurant exhaust appliances*. Bensenville, IL: UL Solutions.

UL. 2015. UL 1479, *Fire tests of penetration firestops*. Bensenville, IL: UL Solutions.

UL. 2017a. UL 199E, *UL LLC outline of investigation for fire testing of sprinklers and water spray nozzles for protection of deep fat fryers*. Bensenville, IL: UL Solutions.

UL. 2017b. UL 705, *Power ventilators*. Bensenville, IL: UL Solutions.

UL. 2017c. UL 8782, *UL LLC outline of investigation for pollution control units for commercial cooking operations*. Bensenville, IL: UL Solutions.

UL. 2019. UL 300, *Fire testing of fire extinguishing systems for protection of commercial cooking equipment*. Bensenville, IL: UL. Solutions.

UL. 2021a. UL 1598, *Luminaires*. Bensenville, IL: UL Solutions.

UL. 2021b. ULC-S645, *Standard for power roof ventilators for commercial and institutional kitchen exhaust systems*. Bensenville, IL: UL Solutions.

UL. 2022a. UL 199, *Standard for safety for automatic sprinklers for fire-protection service*, 13th ed. Bensenville, IL: UL Solutions.

UL. 2022b. UL 508A, *Industrial control panels*. Bensenville, IL: UL Solutions.

USGBC. 2023. LEED rating system. Washington, D.C.: U.S. Green Building Council. www.usgbc.org/leed.

U.S. CPSC. 2023. Protect your family from carbon monoxide poisoning. Bethesda, MD: United States Consumer Product Safety Commission. www.cpsc.gov/safety-education/safety-education-centers/carbon-monoxide-information-center/protect-your-family-from-carbon-monoxide-poisoning--.

VDI. 2022. VDI 2052, *Air conditioning—kitchens*. Düsseldorf-Lohausen: Verein Deutscher Ingenieure.

WCF. 2023. Heat safety. Sandy, UT: WCF Insurance. www.wcf.com/sites/default/files/2020-12/HeatSafetyProgram.docx.

Zhang, J., M. Kong, K.H. Han, B. Guo, and Z. Liu. 2020. *Developing a test method to determine the effectiveness of UVC systems on commercial cooking effluent*. ASHRAE Research Project RP-1614 Final Report. Peachtree Corners, GA: ASHRAE.

Zheng, J., K. Han, B. Guo, Y. Zhang, and C. Wu. 2016. *Countertop commercial appliance emissions*. ASHRAE Research Project RP-1631 Final Report. Peachtree Corners, GA: ASHRAE.

Index